Green Communication with Field-programmable Gate Array for Sustainable Development

The text discusses the designing of field-programmable gate array-based green computing circuits for efficient green communication. This book will help senior undergraduate, graduate students, and academic researchers from diverse engineering domains such as electrical, electronics and communication, and computer.

- Discusses hardware description language coding of green communication computing (GCC) circuits.
- Presents field-programmable gate arrays-based power-efficient models.
- Explores the integrations of universal asynchronous receiver/transmitter and field-programmable gate arrays.
- Covers architecture and programming tools of field-programmable gate arrays.
- Showcases Verilog and VHDL codes for green computing circuits such as finite impulse response filter, parity checker, and packet counter.

The text discusses the designing of energy-efficient network components, using low voltage complementary metal-oxide semiconductors, high-speed transceiver logic, and stub series-terminated logic input/output standards. It showcases how to write Verilog and VHDL codes for green computing circuits, including finite impulse response filter, packet counter, and universal asynchronous receiver-transmitter.

10 HSTL-based FIR filter for GCC   153

11 MOBILE DDR-based FIR filter for GCC   167

12 LVCMOS-based packet counter for GCC   181

13 SSTL-based packet counter of GCC   197

Preface

My co-author, Dr. Pandey, and I share a common research domain of designing the FPGA-based devices for green communication. The RTL coding of all Green Communication circuits is done by Dr. Pandey. I worked on power analysis and simulation of those circuits to make the circuits suitable for Green Communication. In the past recent years, we both are working together on designing power-efficient design on FPGA. The main aspect of writing this book is to let the readers know about the various green communication circuits, various IO standards used with FPGA to optimize the power consumption of the circuits, so that those circuits become part of green communication.

FPGAs are the technology of future. FPGA devices are currently being used in almost all domains, and they are providing better results. FPGAs are used in cybersecurity, artificial intelligence, blockchain, data science, embedded designs, and many more. This book will give the readers a detailed analysis of the green communication circuits with different IO standards. We hope that after reading this book the audience will have detailed knowledge of green communication, FPGA, and will be able to design its own power-efficient circuits that can promote green communication as well as sustainability.

Keshav Kumar

Authors

Prof. Dr. Bishwajeet Pandey is working as Associate Professor in Jain University, Bangalore, India. He has earned his PhD in CSE from Gran Sasso Science Institute, L'Aquila, Italy, under the guidance of Prof Paolo Prinetto, Politecnico Di Torino, Italy. He has worked as an Assistant Professor in the Department of CSE at Birla Institute of Applied Science, India; Assistant Professor in the Department of Research at Chitkara University, India; JRF at South Asian University and Lecturer in Indira Gandhi National Open University. He has completed Master of Technology (IIIT Gwalior) in CSE with specialization in VLSI, Master of Computer Application, R&D Project in CDAC-Noida, India. He has authored and coauthored 150+ papers and these are available on his Scopus Profile (https://www.scopus.com/authid/detail.uri?authorId=57203239026). He has 2000+ citations according to his Google Scholar Profile (https://scholar.google.co.in/citations?user=UZ_8yAMAAAAJ&hl=en). He has experience teaching Cyber Security, Innovation and Startup, Computer Network, Digital Logic, Logic Synthesis, Machine Learning, System Verilog, and so on. His area of research interest is Green Computing, High Performance Computing, Cyber Physical System, IoT, and Cyber Security. He is on the board of directors of many startups of his students, e.g. Gyancity Research Consultancy Pvt Ltd.

Keshav Kumar is currently working as an Assistant Professor at Chandigarh University, Punjab, India. He has earlier worked as a Junior Research Fellow (JRF) with NIT Patna and as an Assistant Lecturer at Chitkara University, Punjab, India. He completed his Master of Engineering in ECE with a specialization in Hardware Security from Chitkara University, Punjab, India. He has published over 30+ research papers in the field of Hardware Security, Green Communication, Low Power VLSI Design, Machine Learning techniques, and IoT. He also has worked with professors of 15 different countries. His area of specialization is Hardware Security, Green Communication, Low Power VLSI Design, Machine Learning techniques, WSN, and IoT. He is also associated with Gyancity Research Consultancy Pvt Ltd. He is a member of IAENG. His Google Scholar profile is Keshav Kumar - Google Scholar and his Scopus Profileis Scopus preview - Kumar, Keshav - Author details - Scopus.

Chapter 1

Introduction to green communication computing (GCC)

LIST OF ABBREVIATIONS

AC	Alternating Current
ASIC	Application-Specific Integrated Circuit
CMOS	Complementary Metal-Oxide Semiconductor
CO_2	Carbon Dioxide
CRT	Cathode-Ray Tube
EPA	Environmental Protection Agency
FPGA	Field Programmable Gate Array
GC	Green Computing
GCC	Green Communication Computing
G. Comm.	Green Communication
GHG	Green House Gases
HDD	Hard Disk Drive
IC	Integrated Circuit
ICT	Information and Communications Technology
IO	Input Output
IT	Information Technology
LAN	Local Area Network
LCD	Liquid Crystal Display
OLED	Organic light-emitting diode
RAN	Radio Access Networks
RTL	Register Transfer Logic
SSD	Solid State Drives

1.1 INTRODUCTION

With the rise in global market value of computing and mobile devices, one cannot deny these devices and peripherals have drastically improved our day-to-day life and work. With the rise of these devices in people, life

DOI: 10.1201/9781003302872-1

1

environment is too facing some problems like deficiency of energy and power. Therefore, in order to promote the ideas of green computing, we must look upon some questions like:

i. Which green computing issues are affecting the environment?
ii. What will be its influence on the environment?
iii. How green computing technologies can improve the environmental conditions?

Therefore, green computing (GC) can be defined as, future-generation environmentally friendly way of utilizing the computers, mobile devices, and their resources. GC is also regarded as green information technology (green IT) [1,2]. In a broad way, the term green computing can also be coined as the method of designing, manufacturing, implementing, using, and disposing the mobile and computing peripherals and devices with the least damage on environment resources.

The terms such as power management of devices, designing of energy-efficient devices, processors and other computer devices, cloud, and virtualization are associated with green computing [3–4]. In cloud and virtualization, we try to communicate the data with cloud server and access the data from cloud. Generally, there are four major points concerned with GC:

i. Green use: It implies minimizing power consumption of computer and mobile devices.
ii. Green disposal: It implies reusing and recycling of unwanted electronic devices.
iii. Green design: It covers the implementation and designing power and energy-efficient devices.
iv. Green manufacturing: Green manufacturing discusses about manufacturing computer and mobile devices with minimized waste.

The communication of green devices with cloud and virtualization is known to be green computing communication (GCC) [5,6]. The overview of GCC is shown in Figure 1.1.

1.2 ORIGIN/HISTORY OF GREEN COMPUTING COMMUNICATION

The idea of GC strikes first in people's mind in the early 1990s. In the era of the early 1990s, the computers and mobiles were quite bulky and power intensive. One cannot imagine how huge amount of power these bulky devices consume to operate. There was a huge amount of power consumption because these devices and peripherals neither have sleep nor hibernate

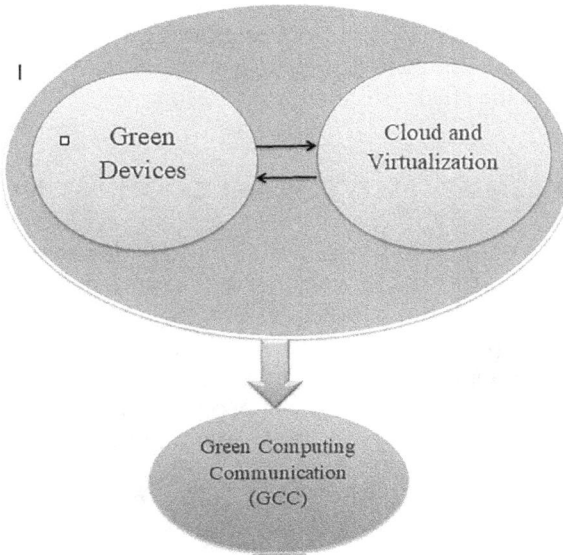

Figure 1.1 Green communications.

option. That means they are consuming power even when they are in idle state. Apparently, nowadays devices have all add-ons. In short, we can say that there was no power- and energy-saving option which allows efficient energy and power management. In view of the above difficulties, U.S. Environmental Protection Agency (EPA) launched a program of Energy Star in 1992 [7].

The program launched by U.S. EPA has laid the basis of designing energy and power-efficient devices for computing. Computers and other computing devices started giving the services like sleep, hibernate, and log off option. These new features in computers allowed to save power when they are of no use. With these features, a huge reduction in energy and power consumption has been observed in computers. Also, this results in reduction of carbon dioxide (CO_2) in the environment.

1.3 GREEN AND SUSTAINABLE COMPUTING

The term sustainability refers to the fact that present generation should meet their essentials without negotiating the demands of future generation. Therefore, green sustainable computing can be quoted as the technology which uses computers and computing facilities without disturbing environmental imbalance. According to the *Kyoto Protocol*, sustainability can be summarized as follows [8,9]:

- To minimize the global reserve extraction to its half
- To dissociate resources' use from economic growth
- To enhance the proficiency of primary resources
- To suppress the energy necessities

1.4 WHY GREEN AND SUSTAINABLE COMPUTING?

Computers and IT disturb the environmental imbalance in several distinguished ways. The effect on environment can be seen from the very first phase, i.e., from production to their utilization and until the disposal of computers and IT. In the present scenario, everyone across the globe needs the mobile and computers in their daily-life process. The use of computers and mobiles saves much precious time of people. The larger number of computing devices not only generates a huge amount of heat but also utilizes a huge amount of power [10–11].

So, what do you think, what will be the effect of these abovementioned consequences on you and on your environment?

The huge amount of heat and enormous power consumption will result in increased amount of CO_2 in the environment. Since the CO_2 is the major cause of greenhouse gases in the environment, it will result in global warming. The increased CO_2 will also affect the ozone layer, and this layer will get thinner and thinner as CO_2 will increase. The CO_2 emitted by the mobile and computing equipment is well described in Figure 1.2.

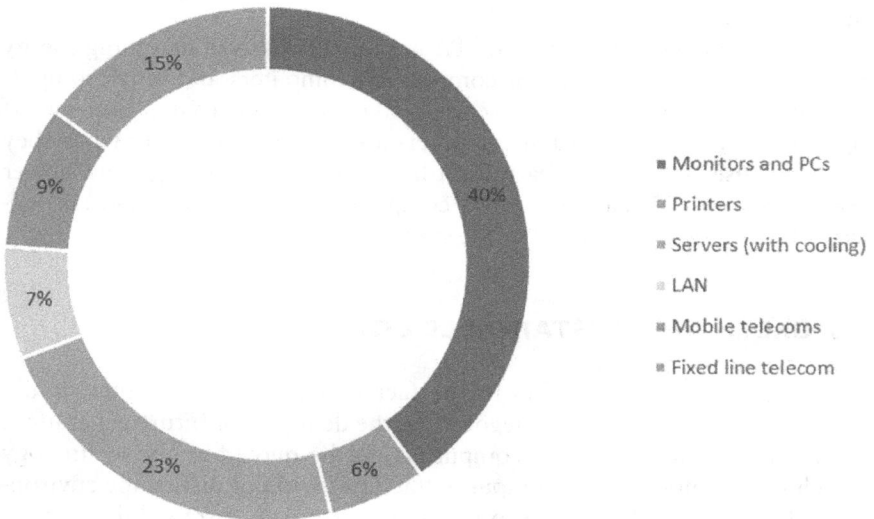

Figure 1.2 CO_2 emission from the mobile and computing equipment.

From Figure 1.4, it is observed the PCs and monitors (40%) come at the top of the table when it comes to CO_2 emission, followed by servers (23%), fixed line telecom (15%), mobile telecoms (9%), LAN (7%), and printers (6%). Sustainable green computing is not all about designing and creating power and energy-efficient system such as hardware equipment and software platforms but also to make these hardware and software to be used in various distinguished IT and communication fields. With all these hardware and software designs, there is a chance of global environmental sustainability.

Considering the abovementioned factors in our mind, we can say that the green sustainable computing infrastructure can help us in the following ways:

- More use of energy- and power-efficient designs
- Promoting the idea of recycling waste equipment
- Use of more environment-friendly things
- Efficient disposal of devices
- CO_2 reduction by using environment-friendly products

1.5 FRAMEWORK OF GREEN COMPUTING

The green computing module is the use of green sustainable devices which consume low power and emit low CO_2 in air. In this module, the main aim of global companies is to become green for environmental sustainability. Therefore, to develop a green company, we should focus on some of the important aspects as follows [12,13]:

a. Company should make and implement plans which support the environment-friendly strategies.
b. Company should focus on the use of recycling.
c. They should try to utilize minimum power and support energy conservation.
d. They should make the computers and IT infrastructure such that their power consumption should be paused when they are not in active state [14].

1.6 GREEN COMMUNICATION (G. COMM.)

1.6.1 Why G. Comm.?

In the current scenario, in the world of wireless and telecommunication, there is drastic problem of emission of greenhouse gases (GHGs) and energy crisis. These two major problems have been a huge barrier in the efficient development of smooth wireless telecommunication. From the recent study, it has been noted that the world of ICT is consuming about 4% of total world's energy and the 4% of the energy is emitting 3% of the total GHG in the environment. As of now, research says that approximately a half of the total global population are mobile and computer users. And as the technology is advancing day by day, the research and the global ICT market predicts that the number of users will increase rapidly. As the user will increase, there will be a huge load on telecommunication network and cloud server to make communication efficient and smooth. The load on telecommunication network will give rise to increased number of mobile networks. The increased mobile network will require more energy for operation and hence will emit more amount of GHG in the environment. The total energy requirement of the IT sector can be examined that about 55% of energy is consumed by the networking and telecommunication infrastructure and 45% of the energy consumption is counted by mobile, computer, and other devices.

Another reason of energy crisis on a global scale other than increased mobile devices and networking infrastructure is rapid increase in world's population. By 2020, the total population of the world is about 7.8 billion, and it is expected to be about 8.55 billion by the end of 2030, 9.74 billion by 2050, and around 10.9 billion by 2100. The global population trend is shown in Table 1.1. With the advancements in medical sciences, it is observed that the lifespan of human has also increased. Around the 1990s, the lifespan was about 64.06 years, and in 2019, it is increased to 72.6 years [15]. Apart from population growth, industrialization and urbanization is also affecting the energy crisis. The cause of urbanization across the globe is change in people's lifestyle and advancement of technologies. Because of this urbanization, the number of industries is also growing across the world and these industries require a lot of energy and power for their operation and also emit a huge amount of CO_2 in air. Excess of CO_2 causes rise of temperature and hence will lead to global warming. According to the UN reports of 2020, it is regarded that 2007 was the first year when the highest number of people moved to urban area from rural areas, and it is estimated that by 2050, about 66% of the world's population will live in urban cities rather than rural areas.

1.7 POPULATION V/S ELECTRICITY AND CO$_2$ CONSUMPTION

The growth of population across the globe leads to further electricity and CO_2 consumption. As per Global Energy Statics 2020, it was found in 2019 that China, USA, India, Russia, and Japan are the top five countries in consuming electricity. These countries consumed 6,510, 3,865-, 1,230-, 922-, and 918-Terawatt hour (TWh) of electricity, respectively, in 2019.

The electrical energy consumed by top ten countries in 2019 is depicted in Table 1.2.

Table 1.1 Population growth
in billion [16]

Years	Population (in billion)
1950	2.54
1955	2.77
1960	3.03
1965	3.34
1970	3.7
1975	4.08
1980	4.46
1985	4.87
1990	5.33
1995	5.74
2000	6.14
2005	6.54
2010	6.96
2015	7.38
2020	7.79
2025	8.18
2030	8.55
2035	8.89
2040	9.2
2045	9.48
2050	9.74
2100	10.9

Table 1.2 Electricity energy consumption
for 2019 [14]

Countries	Electricity consumption (TWh)
China	6,510
USA	3,865
India	1,230
Russia	922
Japan	918
South Korea	553
Canada	543
Brazil	536
Germany	517
France	437

In 2019, the BRICS nations (Brazil, Russia, India, China, and South Africa) only had consumed 78% of the total electricity consumption. The growth of electricity consumption in 2019 was +0.7% which was relatively less when compared with the average in the period of 2000–2018, which was 3% per year. The consumption of electricity in China is increased by 4.5% in 2019. The average in China in the period of 2000–2018 was 10% per year.

1.7.1 CO_2 consumption

There has been a slight decrease in CO_2 emission (−0.2%) in 2019. The USA and Europe had contributed more to reduction in CO_2 emission by −2.4% and −3.9%, respectively. In Asian countries on the whole, there is an increment in CO_2 emission. China shows an increment of +2.8% in 2019 as compared to past years. There has been also reduction in CO_2 emission in India, Japan, and South Korea [14,17]. The top ten CO_2 emitting countries are shown in Table 1.3.

1.7.2 Population v/s total energy consumption across the globe

Since the technology across the globe is getting advanced, the total energy consumption in all countries across the globe is getting increased. According to the statistics of global energy data, it is observed that there is reduction in total energy consumption across the globe 2019. In 2019, the energy consumption increased by +0.6% which is lesser than the average growth for the period of 2000–2018. For the period of 2000–2018, the average growth in energy consumption was +2% per year. Since 2000, the BRICS countries are only consuming +67% of the total world's energy [14].

Table 1.3 CO_2 emission in 2019

Countries	CO_2 emission (Mt CO_2)
China	9,729
USA	4,920
India	2,222
Russia	1,754
Japan	1,045
Germany	673
South Korea	650
Iran	638
Indonesia	581
Canada	569

Table 1.4 Energy consumption of the top ten countries in 2019

Countries	Energy consumption (MTOE)
China	3,284
USA	2,213
India	913
Russia	779
Japan	421
South Korea	298
Germany	296
Canada	295
Brazil	288
Indonesia	269

Here the unit for measuring the energy consumption is millions of tons of oil equivalent (MTOE). It is equal to 4.1868×10^{16} J, or 41.868 petajoules, which is an incredible amount of energy.

The pace of growth of energy consumption in China slowed down than previous years. The growth in 2019 was +3.2% in China. The growth in Russia was +1.8%, and in India, it was only +0.8%. In the USA, Japan, and European Union, the energy consumption has been reduced in 2019 by –1%, –1.6%, and –1.9%, respectively. The top ten countries across the globe with highest amount of energy consumption are: China (3,284), USA (2,213), India (913), Russia (779), Japan (421), South Korea (298), Germany (296), Canada (295), Brazil (288), and Indonesia (269). The energy consumption of the top ten countries is listed in Table 1.4.

1.7.3 What is G Comm.?

Keeping in the minds the demand of energy for the future generation and considering the abovementioned factors came the concept of G. comm. G. comm. can be defined as the practice and promotion of energy-/power-efficient comm. and networking technologies, reducing the use of natural resources in all aspects of comm. system. Therefore, we can say that the future of communication and multimedia with sustainability relies on the G Comm. and its technologies. G Comm. are just like sharing of information, awareness of power and energy, routing, and data transmission which ensures the balanced resource utilization for present as well as upcoming generations. Therefore, G. Comm. is a promising technology for the communication purpose. It is also called as next-generation communication. It cuts down the cost of an organization by minimizing the power consumption [18,19].

1.7.4 Some daily-life examples which ensure the necessity of G Comm.

i. Nowadays, laptops have the power backup of 3–4 hours. But if we replace the standard computing communication devices with the green computing communication devices, then the life of the battery backup will be increased. Therefore, the power backup might extend to +2–3 hours and the laptop won't get much heated up and hence will emit less CO_2.

ii. The standard communication networking devices used in the Wi-Fi routers and communication channels consume huge amounts of power (3–20 W). After replacing these standard communication devices with the green computing devices, the consumption of power will be decreased.

iii. With the help of G Comm. devices, radio access network (RAN) sites can be operated at low electricity consumption. Hence, the efficiency of base stations, core stations, and control sites will be increased.

iv. G. Comm. technologies help in economic growth of a country.

1.7.5 Advantages of GCC

i. GCC means reduced energy and power consumption, hence reduction in GHG and usage of natural resources and fossil fuels.

ii. GCC is a cost-effective mode of communication.

iii. GCC ensures recyclability and reuse of electronics components.

iv. Green ICT utilizes nontoxic components, which reduces hazards related to the health of the user.

v. It helps in improving the economic condition of organizations well as a country.

vi. It helps in reduction of heat from electronic devices [20].

1.7.6 Disadvantages of GCC

i. The cost of implementation is much higher at initial level.

ii. Frequent revolution of technology.

iii. It causes more burdens on an individual [20].

iv. Fewer numbers of books and publications related to GCC.

1.7.7 How can you support the GCC?

We can support the GCC by making some small changes at our end which promotes the GCC technologies [21]. Some major steps taken from our side for promoting the GCC are as follows:

   i. We can start using the products which have Energy Star logo.
  ii. Try to switch off the devices when they are not in use.
 iii. Optimal brightness for the screens, laptops, and TVs reduces power utilization.
 iv. IT products should be purchased from the companies which are environmentally committed.
  v. Instead of discarding the equipment, try to reuse or donate them.
 vi. Print only when required and try to print on both sides of the page.
 vii. Put the computers at hibernate or sleep mode when not using for short time.
viii. Purchase solid state drives (SSD) instead of hard disk drive (HDD). SSD consumes less power in comparison to HDD.
 ix. Try to use cloud servers to store data.
  x. Use email service instead of fax services.
 xi. Use of nonpetroleum-based inks for printers.
 xii. Use of OLED display rather than CRT and LCD displays.

REFERENCES

1. Kurp, Patrick. "Green computing." *Communications of the ACM* 51, no. 10 (2008): 11–13.
2. Harmon, Robert R., and Nora Auseklis. "Sustainable IT services: Assessing the impact of green computing practices." In *PICMET'09–2009 Portland International Conference on Management of Engineering & Technology,* Portland, pp. 1707–1717. IEEE, 2009.
3. Aery, Manish Kumar, and Chet Ram. "Green computing: A study on future computing and energy saving technology." *The International Journal of Engineering Science* 25 (2017). https://www.researchgate.net/publication/323377491_Green_Computing_A_Study_on_Future_Computing_and_Energy_Saving_Technology.
4. The importance of green computing. Onhike. https://onhike.com/the-importance-of-green-computing/135566/. Accessed on 30 June 2023.
5. Raza, Khalid, V. K. Patle, and Sandeep Arya. "A review on green computing for eco-friendly and sustainable it." *Journal of Computational Intelligence and Electronic Systems* 1, no. 1 (2012): 3–16.
6. Khan, Samee Ullah, Lizhe Wang, Laurence T. Yang, and Feng Xia. "Green computing and communications." *Journal of Supercomputing* 63, no. 3 (2013): 637–638.
7. U.S. Environmental Protection Agency | US EPA. https://www.epa.gov/.
8. Sun Microsystems - An overview. ScienceDirect Topics. https://www.sciencedirect.com/topics/computer-science/sun-microsystems. Accessed on 30 June 2023.
9. Sustainable computing. https://computing.fs.cornell.edu/Sustainable/default.cfm. Accessed on 30 June 2023.

10. Naumann, Stefan, Markus Dick, Eva Kern, and Timo Johann. "The greensoft model: A reference model for green and sustainable software and its engineering." *Sustainable Computing: Informatics and Systems* 1, no. 4 (2011): 294–304.

11. Framework of green computing. https://www.researchgate.net/figure/Framework-of-Green-Computing_fig2_323377491. Accessed on 30 June 2023.

12. Jiang, Xiaohong, Han-Chieh Chao, Gabriel-Miro Muntean, George Ghinea, and Changqiao Xu. "Green communication for mobile and wireless networks." *Mobile Information Systems* 2016 (2016): 25–33.

13. Wu, Yongpeng, Fuhui Zhou, Zan Li, Shunqing Zhang, Zheng Chu, and Wolfgang H. Gerstacker. "Green Communication and Networking." *Wireless Communications and Mobile Computing* 2018 (2018). Article ID 1921353. https://www.hindawi.com/journals/wcmc/2018/1921353/.

14. World power consumption | Electricity consumption. Enerdata. https://yearbook.enerdata.net/electricity/electricity-domestic-consumption-data.html. Accessed on 30 June 2023.

15. World population day. United Nations. https://www.un.org/en/observances/world-population-day. Accessed on 30 June 2023.

16. Total data volume worldwide 2010–2025. Statista. https://www.statista.com/statistics/871513/worldwide-data-created/. Accessed on 30 June 2023.

17. Lean ICT: Towards digital sobriety. The Shift Project. https://theshiftproject.org/wp-content/uploads/2019/03/Lean-ICT-Report_The-Shift-Project_2019.pdf. Accessed on 30 June 2023.

18. 26 Key pros & cons of green technologies. E&C. https://environmental-conscience.com/green-technologies-pros-cons/. Accessed on 30 June 2023.

19. Advantages and disadvantages of green technology. RF Wireless World. https://www.rfwireless-world.com/Terminology/Advantages-and-Disadvantages-of-Green-Technology.html. Accessed on 30 June 2023.

20. The advantages of green communication. The Hebs Group. https://hebs-group.co.uk/the-advantages-of-green-communication/. Accessed on 30 June 2023.

21. Mahapatra, Rajarshi, Yogesh Nijsure, Georges Kaddoum, Naveed Ul Hassan, and Chau Yuen. "Energy efficiency tradeoff mechanism towards wireless green communication: A survey." *IEEE Communications Surveys & Tutorials* 18, no. 1 (2015): 686–705.

Chapter 2

Field programmable gate arrays (FPGA)

LIST OF ABBREVIATIONS

AC	Alternating Current
ADC	Analog-to-Digital Converters
ASIC	Application-Specific Integrated Circuit
CLB	Configurable Logic Block
CLPD	Complex Programmable Logic Device
CMOS	Complementary Metal–Oxide–Semiconductor
DAC	Digital-to-Analog Converters
DoD	Department of Defense
EPROM	Erasable Programmable Read-Only Memory
FF	Flip-Flops
FPGA	Field Programmable Gate Array
GC	Green Computing
GCC	Green Communication Computing
G. Comm.	Green Communication
HDD	Hard Disk Drive
IC	Integrated Circuit
ICT	Information and Communications Technology
IO	Input-Output
IT	Information Technology
LAN	Local Area Network
LCD	Liquid Crystal Display
LUT	Look-Up Table
MB	Memory Blocks
MOSFET	Metal–Oxide–Semiconductor Field-Effect Transistor
Muxes	Multiplexers
OLED	Organic Light-Emitting Diode
PLA	Programmable Logic Array
PLD	Programmable Logic Devices
PROM	Programmable Read-Only Memory
RAM	Random Access Memory
RAN	Radio Access Networks

DOI: 10.1201/9781003302872-2

ROM	Read-Only Memory
RTL	Register Transfer Logic
SSD	Solid State Drives
TTL	Transistor–Transistor Logic
VHDL	Very High-Speed Integration Circuit HDL (Hardware Description Language)
VHSIC	Very High-Speed Integrated Circuit

2.1 INTRODUCTION

FPGA stands for field programmable gate array. It is an integrated circuit (IC) which is designed in such a way that it can be reconfigured after the manufacturing by the customer. The term field programmable highlights its characteristics of getting reconfigured, and gate array highlights that millions of logic gates are clustered on a single chip [1].

According to one of the major FPGA manufacturers, FPGA can be defined as a semiconductor device which is composed of a matrix of CLBs (Configurable Logic Blocks), and these CLBs are interconnected via reprogrammable interconnects [2]. The prominent feature of being reprogrammed after its manufacturing distinguishes FPGA from application-specific integrated IC (ASICs), which are manufactured in such a way that it can be programmed only for a specific task.

2.2 HISTORY OF FPGA

Since FPGA is a semiconductor device, its concept started from the arrival of first MOSFET (Metal–Oxide–Semiconductor Field-Effect Transistor) in 1960. But technically, the concept of FPGA developed in people's mind PROM (programmable read-only memory) and PLDs (programmable logic devices). In 1970, the very first PROM was invented. Before the PROM, there had been only the read-only memory (ROM). In ROM, the data was fed at the time of manufacturing only. The first PLA was made in 1975 [3,4]. A PLA is logical array which comprises programmable AND gate and OR gate. Both these PLA and PROM have the feature of getting reprogrammed after the production, i.e., field programmable feature. In 1983, Altera company was founded which introduced the world's first programmable device known as EP300. The first FPGA was invented by **Ross Freeman** in collaboration with Xilinx in 1985. The first FPGA was named XC2064. The very first FPGA XC2064 had 64 CLBs and 3 Look-Up Tables (LUTs).

After the invention of the first FPGA, both Altera and Xilinx ruled the FPGA market from 1985 to 1990. By mid-1990, the other competitors of Altera and Xilinx started to enter the FPGA market. The new competitors in the market reduced the market share of Xilinx and Altera in the early 1990s. In 1993, the Actel (now Microsemi) has the share of 18% in the FPGA industry. In the 1990s, there has been a rapid progress in the FPGA industry. This progress is seen both in mass production and in circuit sophistication. At the early stage of FPGA development, FPGAs were only used for networking and telecommunication purpose. But at the end of the 1990s, the application of FPGA was found in industrial application, automotive use, and consumer use [5–6]. With the more use of FPGA, companies like Microsoft started using the FPGAs in their data centers for their search engine Bing. The use of FPGA has accelerated the performance of Microsoft search engine and reduced their power delivery.

2.3 INTEGRATION

As the technology is getting advanced, the integration of FPGA started with the other embedded microprocessors. The integration leads to the development of a new architecture known as System on Chip (SoCs). The integration and the development of SoC started in 2012. The SoCs are the mirror works of a reconfigurable design of CPU architecture called as SB24. The design was created by *Ron Perloff and Hanan Potash of Burroughs Advanced Systems Group* in 1982. Some examples of this hybrid architecture, also known as SoC, are listed below:

 i. Xilinx Zynq-7000 all Programmable SoC, which comes with a 1.0 GHz dual-core ARM Cortex-A9 processor embedded inside the FPGA's device.
 ii. Altera Arria V FPGA, which has 800 MHz dual-core ARM Cortex-A9 processor.
 iii. The Atmel FPSLIC uses an AVR processor combined with Atmel's programmable architecture.
 iv. Microsemi SmartFusion devices which includes an ARM Cortex-M3 hard processor. This device has a flash memory of 512 Kb and a RAM of 64 Kb. Also it includes analog-to-digital converters (ADC) and digital-to-analog converters (DAC).

The timeline of FPGA industry is described in Table 2.1 and Figure 2.1.

2.4 FPGA ARCHITECTURE

The architecture of the FPGA device is generalized into three types of components which are logic blocks, hard blocks, and clocking components. The architecture is described in Figure 2.2 [7–9].

Table 2.1 Timeline of FPGA industry

Year	Innovation
1960	First MOSFET
1961	First Communication IC
1962	First TTL
1970	PROM
1971	EPROM
1975	PLA
1978	PAL
1983	EEPROM
1985	First FPGA

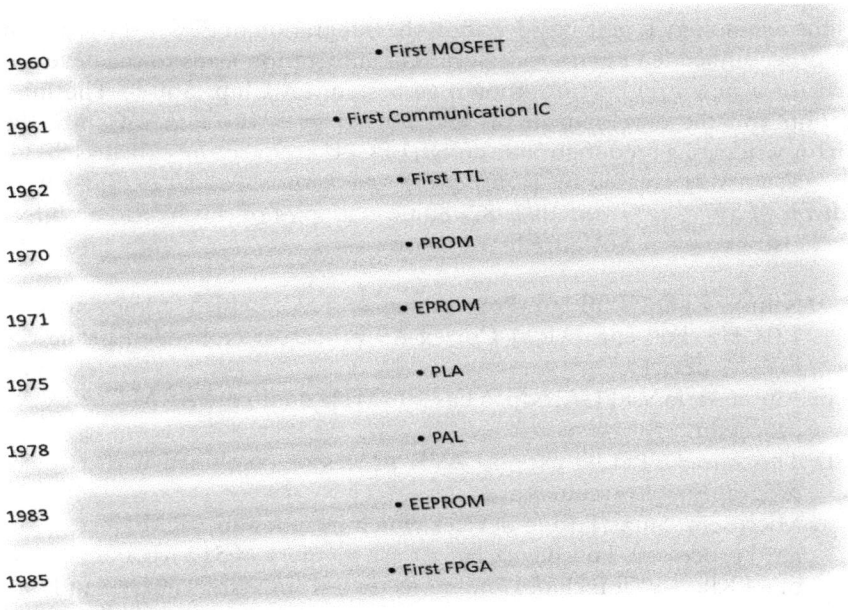

Figure 2.1 Timeline of FPGA industry.

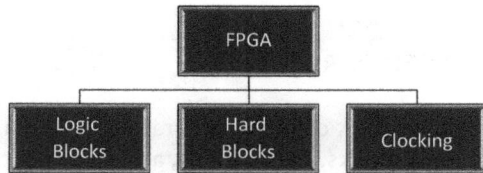

Figure 2.2 FPGA architecture.

The FPGA architecture comprises CLB, LUT, input/output (I/O) pads, buffers (BUFG), memory blocks (MB), flip-flops (FF), multiplexers (Muxes), and random access memory (RAM). The building components of the FPGA device are shown in Figure 2.3.

Logic blocks: It has the components like CLB, LUT, I/O pads, MB, FF, and Muxes.

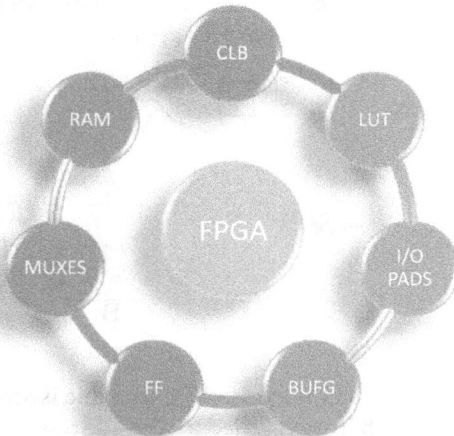

Figure 2.3 Building components of the FPGA.

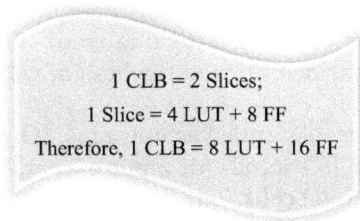

1 CLB = 2 Slices;
1 Slice = 4 LUT + 8 FF
Therefore, 1 CLB = 8 LUT + 16 FF

Figure 2.4 Internal structure of CLB [10].

CLB: It is the basic block of a FPGA device. It allows the user/customer to design and implement any logical function practically. The logical function is being accomplished by the use of a set of two related components known as slices. Inside the CLB, there are two distinguished slices known as SLICEM and SLICEL, or there can be two SLICELs. The slice within the CLB contains LUTs, FFs, and Muxes [9]. The internal structure of CLB is shown in Figure 2.4.

Look-Up Tables (LUTs): As per name, a LUT is a table which gives the output based on the input combination. Let's have an example for a LUT which implements the function of OR gate [11].

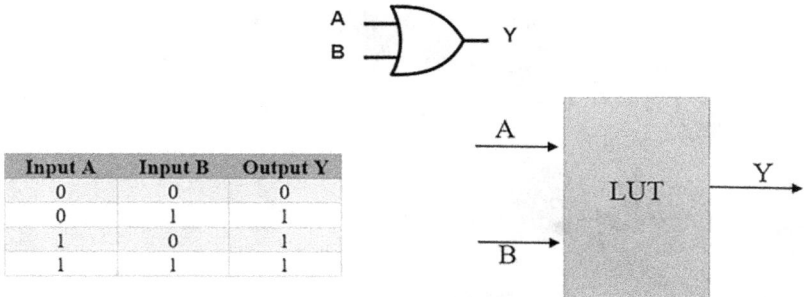

Input A	Input B	Output Y
0	0	0
0	1	1
1	0	1
1	1	1

And now just try to visualize this OR gate table is stored within a RAM. Here the inputs A and B are regarded as the address pins and the Y is data pin. At every change input the output data pin will also be changed as either 0 or 1 depending on the input value.

LUTs are just like our custom-made truth tables which are loaded with values that are applicable to our FPGA design. LUTs inputs and outputs are based on our specific instructions and needs. We can also design our own customized LUT based on our FPGA design. Hence, one can sum up the LUT that LUTs are basically a logic table of various Boolean expression. Instead of using various gates in the design, we use one LUT for that specific design.

I/O pads: I/O pads are also called as user pins, IO pins, IO Blocks (IOB), I/O, and user IOs. Here I/O stands for input-output. Within the FPGA, the IO pins are connected to IO cells. The IO cells get power from VCCIO, i.e., IO power pins. In the FPGA, the user has total control over the IO pins. The IO s can be predefined by the user in FPGA programming. The IOs can be used as either input pins or output pins. Sometimes IOs are also used as bi-directional pins, i.e., tristate buffers. Each FPGA device has a different number of available IO pins. The data of available and used IO pins can be found on datasheets of every FPGA [12].

Flip-flops (FFs): FF are those sequential circuits which are used to save and synchronize the logic between the clock pulse inside the FPGA. With each clock signals, the FF changes its value and holds that value until it receives another clock signal [13]. The symbol of FF is shown in Figure 2.5.

I/P

FF   O/P

ClK

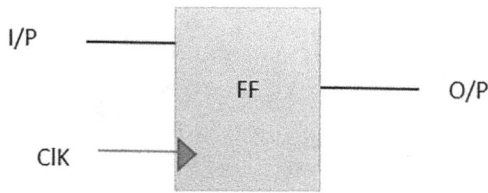

Figure 2.5 Symbol of FF.

Table 2.2 Truth table of D FF

Input D_n	Output Q_{n+1}
0	0
1	1

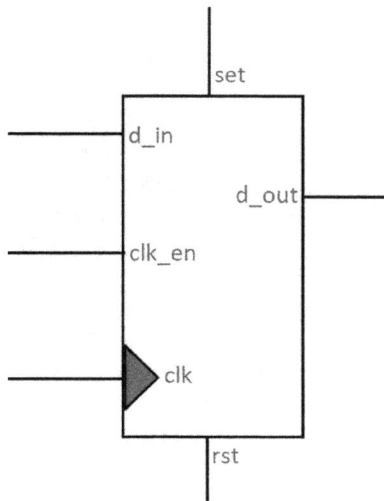

set

d_in

d_out

clk_en

clk

rst

Figure 2.6 FF structure.

There are generally four types of FF in sequential circuit design named as J-K FF, D FF, S-R FF, and T FF. In the FPGA, basically D FF is mostly used in designing any circuit. The truth table of D FF is described in Table 2.2.

The fundamental configuration of the FF contains a clock input (clk), data input (d_in), reset (rst), clock enable(clk_en), and data output (d_out), which is shown in Figure 2.6.

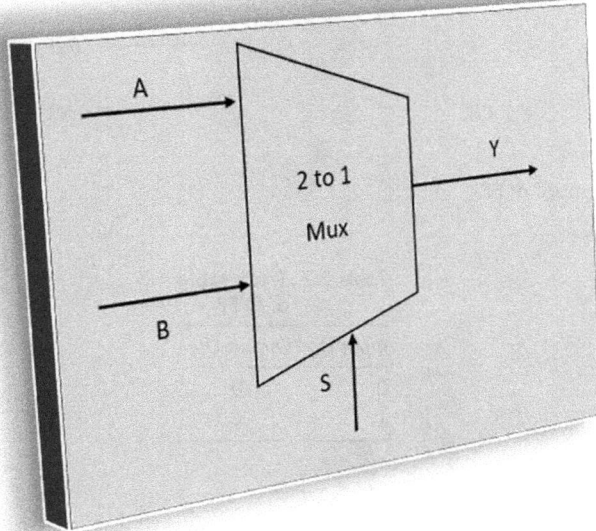

Figure 2.7 Symbol of 2 to 1 Mux.

At the time of regular operation, any value at the d_in pin is latched and passed to the d_out on every pulse of the clk. The intent of the clk_en is to allow the FF to hold a certain value for more than one clk signal. New input is only latched and passed to the output when both clk and clk_en are equal to each other.

Multiplexers (Muxes): A multiplexer (Mux) is an alternative word for a selector. It works just like a railroad switch. The railroad switch operates via some external control which train gets to link to the destination track. The exact identical theory is used with a 2 to 1 Mux. Two inputs can link to a single output.

In a FPGA design, Muxes are used every time with different sizes and various configurations. The symbolic image of 2 to 1 Mux is shown in Figure 2.7.

The output expression for the 2 to 1 Mux is given as:

$$Y = SA + SB$$

The truth table which is fed in LUT for 2 to 1 mux is shown in Table 2.3.

Here the inputs for the Mux are *A* and *B*, *Y* is the output for the Mux, and *S* is the control/select signal. Muxes come in all possible combinations, depending on our used case. Typically, some of the inputs are selected for a single output. But the reverse can also be true and it will still be a Mux. A single input could be selected to any number of outputs [14].

Table 2.3 Truth table for
2 to 1 Mux

S	A	B	Y
0	0	0	0
0	0	1	1
1	1	0	1
1	1	1	1

Memory block (MB): The FPGA device has embedded memory components which can be used as RAM, shift register, or ROM. LUTs and Block RAM (BRAM) are also used as FPGA memory [15].

Hard blocks: The advanced FPGA which has some extra components other than the abovementioned components lies under this category. Examples of such components include multipliers, embedded processors, DSP blocks, and embedded memories. The high-end FPGA includes IP core as processor, Wi-Fi modules, Ethernet point, HDMI port, and USB ports. These components clustered on FPGA device are not built along with LUTs. These are some extra components which help in enhancing the performance of the FPGA and help in utilizing blow amount of power [16,17].

Clocking: Every sequential design built in FPGA requires a Clock (clk) pulse. The clk signal is used to determine how rapidly the FF (or a group of FF) runs. The clk signals are activated according to the frequency of clk. The various clk signals associated with the FPGA design allow different areas of FPGA to be operated at various speeds.

Clock domain in FPGA: The FPGA device has internal PLL (phase locked loop) which helps in generating various clk signals of different frequency. The responsibility of the clk signal is to drive the FPGA design and determine the speed of processing the data bits. Faster the clk speed, faster the processing of data. But a design always doesn't require faster clk signal. A FPGA design built with clk domain is shown in Figure 2.8.

Types of FPGA: There are various types of FPGAs available in the market. Depending upon the programming technologies, the FPGAs can be categorized as follows:

i. SRAM-Based FPGAs: In the semiconductor world, there are generally two types of RAM used: DRAM (i.e., dynamic RAM) and SRAM (i.e., static RAM). SRAM is based upon the transmission gates, transistors, and multiplexers. SRAM FPGA can be reprogrammed, but their stored data is lost when the power is turned off. The programming done in these FPGA cannot be recovered when the power is again supplied. In order to get rid of this problem, these FPGAs need an external memory to store the program [18].

FPGA

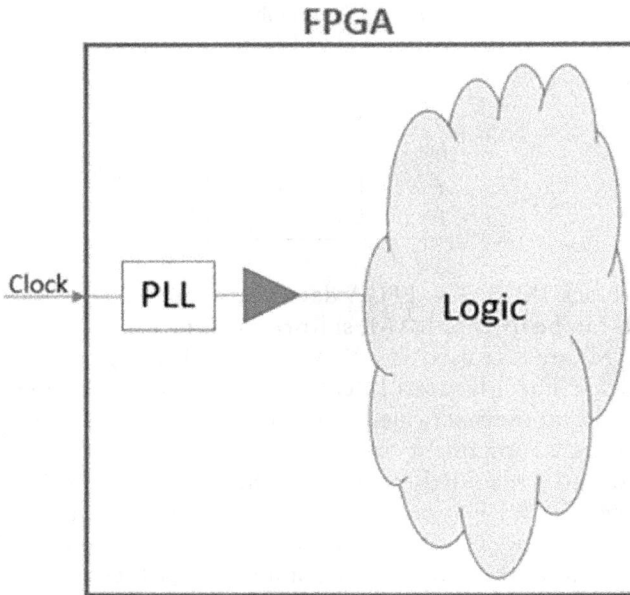

Figure 2.8 A FPGA design built with clk domain.

ii. Flash-Based FPGA: These FPGAs use floating gate cells as switches which improve the area efficiency of the device. The data with these FPGAs is not lost when the device power is switched off. These FPGAs do not require any external memory to store the data, and they also cannot be reprogrammed unlimited times because of charge build-up in the oxide [18].

iii. Anti-Fuse FPGA: These FPGAs are designed up with an anti-fuse CMOS technology. The unique characteristics of these FPGAs is that these devices cannot be reprogrammed. The program stored in the FPGA is not lost when the power is switched off [18].

Differences: The differences between the above-based three types of FPGAs are described in Table 2.4.

FPGA programming: This section will cover the methods and languages used to program a FPGA device. There are in general two languages used to program a FPGA device. These languages are VHDL and Verilog.

VHDL: VHDL is the acronym for VHSIC HDL, i.e., Very High-Speed Integrated Circuit Hardware Description Language. This language is used to program FPGAs and ICs. This language can also be used as a general-purpose parallel programming language. Department of Defence (DoD) introduced the VHDL language in 1981, under the VHSIC module [19].

Table 2.4 Differences between the above-based three types of FPGAs

Characteristics	SRAM	FLASH	Anti-fuse
Power Consumption	High	Low	Low
Reconfigurable	Yes	Yes	No
Actual Speed	Fast	Slow	Fast
Theoretical Speed	Slow	Slow	Fast
Volatility	Volatile	Non-volatile	Non-volatile
Density	Less Dense	Between-SRAM and anti-fuse	Denser
Security	Low	High	High

Table 2.5 Development history of VHDL language

Year	Works
1980	DoD desired to make circuit design self-documenting
1981	Introduction of VHDL under VHSIC
1983	The development of VHDL began with a joint effort by IBM, Inter-metrics, and Texas Instrument
1985	VHDL Version 7.2
1987	VHDL became IEEE Standard 1076-1987
1993	VHDL was re-standardized
1996	A VHDL package used with synthesis tools and become a part of the IEEE 1076 standard
1999	Analog Mixed Signal extension
2008	IEEE Standard 1076-2008

VHDL language is standardized by Institute of Electrical and Electronics Engineers (IEEE) standards. The development history of VHDL language is shown in Table 2.5.

There are several purposes of using VHDL language, and some are listed below:

i. It is used for describing the hardware model.
ii. It is used for simulating the hardware design on FPGA board.
iii. It is used for synthesis purpose.
iv. It is used to describe the hardware architecture.

2.5 BASIC ELEMENTS FOR VHDL PROGRAMMING

To execute a VHDL program on any simulator, there are three elements that must be configured in VHDL programming.

    i. Entity: It is used to define the port in the design. The port may be input, output, inout, or buffer port. The syntax to state entity is as:
        entity name_of_entity is
        Port declaration;
        end name_of_entity;

   ii. Architecture: The following command is used to define the architecture of VHDL language.
        architecture name_architecture of name_of_entity is
        begin
        (program statements)
        end name_architecture;
        Example:
        architecture synthesis of xor is
        begin
         z <= x XOR y;
        end synthesis;

  iii. Configuration: It defines how the design order is organized. It can also be used to associate architecture with an entity.

Verilog: It is an HDL which is used to model a digital design such as FF, networks, microprocessors, and memory element. Digital circuits are very easy to design with Verilog HDL, and also their debugging is quite easy. It is the most used HDL in semiconductor industry [20]. Verilog allows the design engineer to make their design either by top-down or bottom-up methodology.

- Top-down methodology: It enables early testing, quick change of various technology, and the design of standardized structures and provides several other advantages.
- Bottom-up methodology: Each layout is carried out using the standard gates at the gate level. This architecture provides a way for modern systemic, hierarchical methods of design to be designed.

History of verilog: *Prabhu Goel, Phil Moorby, Chi-Lai Huang, and Douglas Warmke* were the great men who were the creator of the Verilog HDL. The Verilog HDL was created in the late 1983 and in starting of 1984. Gateway Design Automation was the organization which developed the Verilog HDL [21]. The detailed history of Verilog HDL is described in Table 2.6.

Requisites: There are some basic points to be required to learn the Verilog HDL:

Table 2.6 History of Verilog

Year	Development
1983 and 1984	Creation of Verilog HDL
1989	Gateway Design Automation acquired by Cadence Designs Systems
1990	Verilog came in public domain
1995	Verilog got IEEE standards 1364-1995
2001	IEEE 1364-2001
2005	System Verilog IEEE 1364-2005
2009	System Verilog IEEE 1800-2009
2017	System Verilog IEEE 1800-2017

Table 2.7 Verilog and VHDL differences

Verilog	VHDL
Multi-dimensional array is not supported	Multi-dimensional array is supported
Mod operator is not available	Mod operator is available
It is easy to learn	It is difficult to learn
Defining data types is not allowed	Defining data types is allowed
It does not allow concurrent calls	It allows concurrent calls
Unary reduction operator is available	Unary reduction operator is not available

- One should be familiar with logic gate, Boolean algebra, Combinational and Sequential Circuits, operators, and other principles.
- One should know about ASIC and FPGA simulation and synthesis.
- One should have the knowledge of timing analysis, path delay, and clk timings.

Difference between Verilog and VHDL: The difference between the two HDL is described in Table 2.7.

2.5.1 Tools for FPGA programming

To program a FPGA device, one should have the knowledge of simulators and tool. A FPGA device can only be programmed with the help of a simulator. There are various simulators available in the markets, some commonly used simulators for FPGA are such as Xilinx and Vivado ISE design suite owned by Xilinx and Modelsim-owned Mentor Graphics.

The major commercial simulators used for FPGA programming are described in Table 2.8.

Table 2.8 FPGA simulators [22]

S. No.	Name of simulator	Company	Description
i.	Active-HDL/ Riviera-PRO	Aldec	This simulator provides wide-ranging design environment to the FPGA applications. Active-HDL are licensed by Aldec organization to Lattice semiconductor. FPGAs of Lattice semiconductor are found in this simulator. This is a low-price simulator owned by Aldec. Aldec provides high-performance and expensive simulator too, that is known as Riviera-PRO. Its aim is to use advanced verification methodologies including assertion-based verification and UVM to verify large FPGA and ASIC devices.
ii.	Incisive Enterprise Simulator	Cadence Design Systems	Cadence developed its own simulator, called as NC-Verilog. The modern version of the NCsim family, known as Incisive Enterprise Simulator, which supports Verilog, VHDL, and System Verilog support.
iii.	ISE Simulator	Xilinx	ISE Simulator owned by Xilinx comes with ISE Design Suite. It supports mixed language mode. FPGAs of different families such as Virtex, Spartan, Artix, and Kintex are found in this. It also gives supports to CLPDs.
iv.	Aeolus-DS	Huada Empyrean Software Co., Ltd	It supports pure Verilog simulation.
v.	CVC	Tachyon Design Automation	This simulator supports only Verilog language.
vi.	Metrics Cloud Simulator	Metrics Technologies	This is a cloud platform to use system Verilog. It has all the essential FPGAs listed. It also uses (Application Programming Interface,) APIs for evaluation.
vii.	PureSpeed	Frontline	This was the very first simulator which can run on Windows OS. In 1998 the simulator was sold to Avant!, and in 2002 it was take over by Synopsys. But later Synopsys also discontinued it.
viii.	Modelsim and Questasim	Mentor Graphics	Both are the widely used HDL simulators. Mentor Graphics launched Questasim in 2005, which provides high-performance System Verilog and Mixed simulator. It also supports OVM and UVM methodologies.
ix.	MPSim	Axiom Design Automation	This simulator provides fast simulation to Verilog, System C, and System Verilog language.
x.	Quartus II Simulator (Qsim)	Altera	This simulator is owned by Altera which supports System Verilog, VHDL, and AHDL.

(Continued)

Table 2.8 (Continued) FPGA simulators [22]

S. No.	Name of simulator	Company	Description
xi.	VCS	Synopsys	VCS is owned by Synopsys which supports the following languages such as: VHDL, Verilog, SystemVerilog, Verilog AMS, SystemC, and C/C++.
xii.	Xilinx and Vivado	Xilinx	This simulator supports all the FPGAs developed by Xilnix and provides simulation of both VHDL, Verilog, and mixed language support.

Table 2.9 FPGA vs ASIC

S. No.	FPGA	ASIC
i.	Reconfigurable device. This can be reprogrammed after post manufacturing. Also, it has a feature of getting reconfigured as a part instead of whole circuit.	Permanent circuitry. ASIC work as the same for the entire life, as it has been programmed at the time of manufacturing.
ii.	VHDL or Verilog is used to program a FPGA.	Programming is the same as FPGA.
iii.	Starting cost is cheap. The cost is around 30 USD.	The cost of starting with an ASIC is very high around of millions of USD.
iv.	Mass production is avoided.	Suited for volume mass production.
v.	Energy efficiency is lesser than ASIC.	More energy efficient as compared to FPGAs.
vi.	Operating frequency is limited as compared to ASIC	Operating frequency is much higher than FPGAs.
vii.	Designing analog device is not possible.	Analog design can be performed.
viii.	Generally preferred for prototyping design.	Not suited for prototyping.

2.5.2 Differences between FPGA, ASIC, CLPD, and microcontrollers

I. FPGA vs ASIC, described in Table 2.9.
II. Visual comparison of FPGA and ASIC, described in Table 2.10.
III. FPGA vs CLPD, described in Table 2.11.
IV. FPGA vs Microprocessor, described in Table 2.12.

2.5.3 FPGA manufacturer/vendors

The major manufacturers of FPGA are Xilinx (now acquired by AMD) and Altera (now subsidary of Intel). These two companies are the market leaders which control about 90% of the FPGA market. Both of these two companies provide software and FPGA which supports the Windows and Linux environment. It allows the user to test, implement, simulate, and synthesis its program on software and hardware [23].

Table 2.10 Visual comparison

	FPGA	ASIC
NRE	✓	
Performance		✓
Time to market	✓	
Design flow	✓	
Per unit cost		✓
Entry barrier	✓	
Energy efficient		✓
Analog blocks		✓

Table 2.11 FPGA vs CLPD

S. No.	FPGA	CLPD
i.	FPGAs are more configurable in comparison to CLPD	Less configurable to FPGA
ii.	SRAM-based FPGAs are volatile	CLPDs are non-volatile
iii.	Complicated timing analysis	Deterministic timing analysis
iv.	Higher power idle power consumption	Lower idle power consumption
v.	More expensive for the use	Cheaper than FPGAs
vi.	Massive built in logic resources	Low built-in logic resources
vii.	FPGAs have DSP, IP, and UART blocks	Don't have these blocks

Table 2.12 FPGA vs microprocessor

S. No.	FPGA	Microprocessor
i.	Provides parallel processing and can execute multiple instruction	Sequential processing and allows only one instruction at one time
ii.	Not useful for serial communication	Useful for serial communication
iii.	FPGA devices are costly	They are of low cost
iv.	Can be used as microprocessor	Cannot be used as FPGA
v.	Consumes more power	Power consumption is low
vi.	Provides high throughput	Provides low throughput

Xilinx manufactures two types of FPGAs:

i. FPGA with (central processing unit) CPU
ii. FPGA without CPU

i. FPGA with CPU: There are three different FPGA of Zynq series which comes under this category.
   a. Zynq 7000 series:
      o Launched in 2011
      o Fabrication process (gate size) 28 nano meter (nm)
      o Integrated with dual or single ARM Cortex A9 CPU
   b. Zynq Ultra Scale +:
      o Launched in 2015
      o Fabrication process (gate size) 16 nm
   c. Versal:
      o Announced by Xilinx in 2018
      o Versal chips contain FPGA components plus GPU, CPU, and DSP block too
      o Process technology is 7nm
ii. FPGA without CPU: Examples are such as Virtex, Spartan, Artix, and Kintex FPGAs

Altera has launched three distinguished versions of FPGAs such as:

i. High-range FPGA: Stratix Series
ii. Mid-range FPGA: Arria Series
iii. Low-cost FPGA: Cyclone series

The other vendors of FPGA are Microchip, Lattice Semiconductor, Quick Logic, Achronix, and Tabula.
Applications: The application of FPGA is seen almost in every industry. The major industries which use the FPGA are listed below:

i. Data centers
   o Security
   o Switches
   o Routers
   o Servers
   o Gateways
ii. Aerospace and defense
   o Communication
   o Avionics
   o Aviation
   o Secure solution

iii. Security
  o Hardware security implementation
  o Image processing
  o Key compromising
  o Hardware module foe security algorithm
iv. Video and audio processing
  o For connectivity
  o For DSP purpose
  o Speech recognition
  o Software-defined ratio (SDR)
  o High resolution video
  o Video over IP
  o Computer vision
v. Scientific instruments
  o Phase locked loops (PLL)
  o Radio astronomy
  o Lock-in amplifiers
vi. Medical
  o CT scan
  o MRI
  o Ultrasound
  o X-ray
  o Surgical systems
vii. Wireless communication
  o Data transmission
  o Network processing
  o Mobile backhauls
  o For connectivity
viii. Consumer electronics
  o Digital display
  o Digital camera
  o Set-top boxes
  o Printers
  o Portable electronics devices

2.5.4 Market contribution

Today, "Xilinx is the world's leading supplier of all programmable FPGAs, SoCs, and 3D ICs," according to the company. From programmable logic to programmable systems integration, these industry-leading devices are combined with a next-generation design environment and IP to support a wide variety of customer needs. Aerospace/Defense, Automotive, Broadcast, Consumer, High-Performance Computing, Industrial/Medical, Wired, and Wireless are among the end markets served by Xilinx [24]. Xilinx currently has about 3,000 staff, 20,000 clients, 2,500 patents, and a market share of more than 50% ($2.2 billion) of the $4 billion programmable market. Altera ($1.8 billion), Actel (now part of Microsemi), and Lattice ($300 million) are the other prominent fabless programmable firms. Achronix and Tabula, both newcomers to the FPGA industry, will be among Intel's first fab customers at 22 nm [25]. In 2010, FPGA market size was estimated around 2.75 billion USD. In 2013, the market size was increased by 96.36%, and it became 5.4 billion USD. In 2020, it became 9.8 billion USD. The increment was 256.36% as compared to 2010 and 81.48% as compared to 2013. And by 2027, it is estimated that the market size will increase by 91.83% as compared to 2020 and the FPGA market will be of 18.8 billion USD.

REFERENCES

1. Hauck, Scott, and Andre DeHon. *Reconfigurable computing: The theory and practice of FPGA-based computation.* Elsevier, Dordrecht, 2010.
2. Llamas, L.. ¿ Qué es una FPGA? Motivos de su auge en la comunidad Maker. Ingeniería, informática y diseño, 2017.[En línea:] http://bit. ly/2XKOqPX..
3. Yang, Hai-gang, Jia-bin Sun, and Wei Wang. "An overview to FPGA device design technologies." 电子与信息学报 32, no. 3 (2010): 714–727.
4. Field programmable gate array (FPGA) history. (2023). HardwareBee. https://hardwarebee.com/field-programmable-gate-array-fpga-history-applications/#:~:text=In%20the%20late%201980s%2C%20the,patented%20the%20creation%20in%201992.
5. Shirriff, Ken. Reverse-engineering the First FPGA chip Xilinx XC 2064. (2020). SemiWiki. https://semiwiki.com/fpga/290990-reverse-engineering-the-first-fpga-chip-xilinx-xc2064/.
6. Amano, Hideharu, ed. *Principles and structures of FPGAs.* Springer, Dordrecht, 2018.
7. Kumar, Keshav, K. R. Ramkumar, and Amanpreet Kaur. "A lightweight AES algorithm implementation for encrypting voice messages using field programmable gate arrays." *Journal of King Saud University-Computer and Information Sciences* 34, no. 6, Part B (June 2022): 3878–3885.
8. Kumar, Keshav, K. R. Ramkumar, and Amanpreet Kaur. "A design implementation and comparative analysis of advanced encryption standard (AES) algorithm on FPGA." In *2020 8th International Conference on Reliability, Infocom Technologies and Optimization (Trends and Future Directions) (ICRITO)*, Noida, pp. 182–185. IEEE, 2020.
9. Pandey, Bishwajeet, Keshav Kumar, Sri Chusri Haryanti, Rajina R. Mohamed, and D. M. Akbar Hussian. "Power efficient control unit for green communication." *Test Engineering & Management* 83 (March/April 2020): 13422–13427.
10. Roshanmaharana. Introduction to FPGA with Verilog. (2023). Medium. https://medium.com/@roshanmaharana1510/introduction-to-fpga-with-verilog-ab6f02cbda34.

11. Pandey, Bishwajeet, Keshav Kumar, Aiza Batool, and Shabeer Ahmad. "Implementation of power-efficient control unit on ultra-scale FPGA for green communication." *3c Tecnología: glosas de innovación aplicadas a la pyme* 10, no. 1 (2021): 93–105.
12. Haripriya, D., Keshav Kumar, Anurag Shrivastava, Hamza Mohammed Ridha Al-Khafaji, Vishal Moyal, and Sitesh Kumar Singh. "Energy-efficient UART design on FPGA using dynamic voltage scaling for green communication in industrial sector." *Wireless Communications and Mobile Computing* (2022): 1–9. Article ID 4336647. https://doi.org/10.1155/2022/4336647.
13. Kumar, Keshav, Bishwajeet Pandey, Amit Kant Pandit, Y. A. Baker El-Ebiary, Salameh A. Mjlae, and Samer Bamansoor. "Design of low power transceiver on Spartan-3 and Spartan-6 FPGA." *International Journal of Innovative Technology and Exploring Engineering* 8, no. 12S2 (2019): 27–30.
14. Types of rail switches and features of different types of rail switches. (2023). http://www.rail-fastener.com/types-of-rail-switches.html.
15. Kumar, Tanesh, Bishwajeet Pandey, Teerath Das, and B. S. Chowdhry. "Mobile DDR IO standard based high performance energy efficient portable ALU design on FPGA." *Wireless Personal Communications* 76, no. 3 (2014): 569–578.
16. Pandey, Bishwajeet, and Manisha Pattanaik. "Clock gating aware low power ALU design and implementation on FPGA." *International Journal of Future Computer and Communication* 2, no. 5 (2013): 461.
17. The ultimate guide to FPGA architecture. (2023). HardwareBee. https://hardwarebee.com/the-ultimate-guide-to-fpga-architecture/.
18. Zeidman, Bob. All about FPGAs. (2006). EETimes. https://www.eetimes.com/all-about-fpgas/.
19. VHDL tutorial. (2023). JavaTpoint. https://www.javatpoint.com/vhdl.
20. VLSI design - Verilog introduction. (2023). Tutorials Point. https://www.tutorialspoint.com/vlsi_design/vlsi_design_verilog_introduction.htm.
21. Tala, Deepak Kumar. Verilog tutorial. (2003). ASIC World. http://www.asic-world.com/verilog/veritut.html (in English).
22. List of HDL simulators. (2023). Wikipedia. https://en.wikipedia.org/wiki/List_of_HDL_simulators.
23. FPGA market size, share and trends forecast to 2026. (2023). Markets andMarkets™. https://www.marketsandmarkets.com/Market-Reports/fpga-market-194123367.html#:~:text=The%20major%20companies%20in%20the,and%20QuickLogic%20Corporation%20(US).
24. What is an FPGA? Field programmable gate array. (2023). Xilinx. https://www.xilinx.com/products/silicon-devices/fpga/what-is-an-fpga.html.
25. FPGA market will touch USD 25.85 billion at an 11.2% CAGR. (2023). GlobeNewswire. https://www.globenewswire.com/en/news-release/2022/04/21/2426666/0/en/FPGA-Market-will-Touch-USD-25-85-Billion-at-an-11-2-CAGR-by-2030-Report-by-Market-Research-Future-MRFR.html.

Chapter 3

HDL coding of GCC Circuits

LIST OF ABBREVIATIONS

AC	Alternating Current
ADC	Analog-to-Digital Converters
ASIC	Application-Specific Integrated Circuit
CLB	Configurable Logic Block
CLPD	Complex Programmable Logic Device
CMOS	Complementary Metal–Oxide–Semiconductor
CPU	Central Processing Unit
DAC	Digital-to-Analog Converters
DoD	Department of Defense
EPROM	Erasable Programmable Read-Only Memory
FF	Flip-Flops
FIR Filter	Finite Impulse Response Filter
FPGA	Field Programmable Gate Array
GC	Green Computing
GCC	Green Communication Computing
G. Comm.	Green Communication
HDD	Hard Disk Drive
IC	Integrated Circuit
ICT	Information and Communications Technology
IO	Input Output
IT	Information Technology
LAN	Local Area Network
LCD	Liquid Crystal Display
LUT	Look Up Table
MB	Memory Blocks
MOSFET	Metal–Oxide–Semiconductor Field-Effect Transistor
Muxes	Multiplexers
OLED	Organic Light-Emitting Diode
PLA	Programmable Logic Array

DOI: 10.1201/9781003302872-3

PLD	Programmable Logic Devices
PROM	Programmable Read-Only Memory
RAM	Random Access Memory
RAN	Radio Access Networks
ROM	Read-Only Memory
RTL	Register Transfer Logic
SSD	Solid State Drives
TTL	Transistor–Transistor Logic
UART	Universal Asynchronous Receiver Transmitter
VHDL	Very High-Speed Integration Circuit HDL (Hardware Description Language)
VHSIC	Very High-Speed Integrated Circuit

3.1 WHAT IS HDL?

Digital circuits are generally composed of transistors interconnected to each other. With the help of a hierarchical system, we develop and study these circuits. Moreover, we might have a view of the central processing unit (CPU) as a large ocean of transistors, but the organization of transistors into logical gates, registers, and memory units is quite handy and simpler.

This hierarchical structure enables one to view a digital circuit efficiently by interconnected diagrams, which is known as schematic. This visual approach is intuitive but becomes tedious with the difficulty. The use of a textual vocabulary that is deliberately designed to capture explicitly and concisely the distinguishing characteristics of digital architecture is another means of describing digital circuits. The language which does the same is known as hardware description language (HDL) [1].

A textual description of operators, expressions, sentences, inputs, and outputs is provided by HDL. The HDL compilers provide a gateway map rather than creating a machine executable format. The obtained gate map is then downloaded to the programming computer to verify the circuit operations. It can be defined as a structural, behavioral, and door level in any digital circuit, which can be considered an excellent CPLD, FPGAs, and ASICs programming language. Verilog, VHDL, and SystemC are the three popular HDLs. System C is the latest of these. Verilog and VHDL are frequent in most sectors. Verilog, standardized as IEEE 1364, is one of the major hardware description languages for all circuit's architecture. It includes components that allow for dataflow, behavioral, and structural description. VHDL is IEEE 1164 standard. The concept is made up of multiarchitectural entities. SystemC is a language consisting of a collection of C++ and macros. It enables the degree and transaction modeling of the electronic structure [2,3].

3.2 WHY DO WE NEED HDLS?

Dramatic changes have started to arise in the IC world with the Moore law of 1970. The engineers have made this change to produce complicated digital and electronic circuits. But the problem was the lack of a better language that allows the co-design of hardware and software. Complex architectures for digital circuits take more time to create, synthesize, simulate, and debug. The introduction of HDLs helped to address this issue by enabling a different team to operate each module [4].

The introduction of the HDL helped designers and researchers to identify all the parameters, such as power, performance, delay, latency, test coverage, functionality, and area consumption, needed for design. The designer will then conclude the requisite engineering agreements and refine the concept more effectively. The description of electronic circuits often requires a simple grammar, expressions, sentences, concomitant, and sequential programming. The HDL can be used to achieve all these functions. Now, the only distinction is that HDL includes the time details of a design when contrasting HDL and C languages.

3.3 DESIGN AND STRUCTURE OF HDL

In HDLs, the layout of any electronic circuit and IC is a normal text-based expression. Unlike other programming languages, however, HDLs often contain an explicit notion of time that is an important hardware attribute. HDL uses text and netlist language design, the computer-aided design of electronic circuits. HDL can be used to express designs in behavioral, structural, or register-transfer-level architectures for the same circuit functionality. It can be used for two cases: one is synthetization and the other is simulation that decides the architecture and logic gate layout. HDL provides an executable file for hardware design. The hardware designer can model a part of the hardware before actual creation by using a program programmed to execute the fundamental schematic of language statements and simulate it with time. Simulators are available and can support the discrete (digital) and continuous (analog) modeling [5,6].

3.4 DESIGN

HDL provides great efficiency; therefore, a majority of modern digital circuit designs revolve around HDL. Most designs start as a collection of specifications or an architectural diagram at high levels. In flowchart applications, control and decision systems are also prototyped or input to a state diagram editor. The way the HDL definition is written depends heavily on the type of the circuit and the designer's choice for coding. The HDL is simply the capture language that starts with an algorithmic definition of the highest level, such as a mathematical C++ model. Languages like Perl are also used by designers to create repetitive circuit structures in HDL language. Special text publishers provide features to allow automatic indentation, color syntax-dependent, and entity/architecture/signal declaration macro-based expansion.

The HDL description is subjected to a series of automated tests in preparation for synthesis. The control deviations from uniform code instructions are recorded, and possible ambiguous code constructions are identified before misinterpretation and logical errors, such as floating ports or shortened outputs, can be verified. This method helps to solve mistakes before synthesizing the code [7].

HDL architecture usually finishes in the phase of synthesis in industry talk. The netlist is then passed on to the back-end stage when the synthesis tool maps the HDL definition to a portal netlist. HDLs can, or cannot, play an important role in a back-end flow, depending on the physical technology (FPGA, ASIC gate array, and ASIC standard cell). In general, with a physically realizable design stream, the design database is gradually loaded with technological information that cannot be contained in a general HDL definition. Finally, a built-in circuit is produced or configured for use.

3.5 HDL VS PROGRAMMING LANGUAGES

When we try to learn HDL, we are already acquainted with programming languages. HDLs look like high programming languages such as C and Python, but the fundamental difference is that statements in HDL are parallel, while programming languages describe a sequence. We know that the processor runs lines of codes one at a time by writing a computer program

Table 3.1 Differences between HDL and programming languages

Programming languages	HDL
Programming languages are used to write a set of codes/instructions to allow a CPU to perform a specific task.	HDL describes the behavior and architecture of electronic circuits and mostly, digital logic circuits.
Used to develop many applications and tools.	Only used for designing digital circuits.
Easy to cope with programming languages.	Good knowledge of digital electronics is required to write an HDL code.
Suitable with all processors.	Compatible with only FPGAs and ASIC.
Examples: C, C#, C++, Java, and R	Examples: Verilog and VHDL.

or firmware module according to the top-to-bottom structure [8]. We describe the digital designs in HDL language, which can be used simultaneously by different parts of this hardware, although the corresponding codes are written in top-to-bottom design. Differences between these are described in Table 3.1.

3.6 GREEN COMPUTING (GC) CIRCUITS (GCC)

In recent years, the idea of green computing has become more prominent across the globe, because many companies have started to look at their carbon emissions and their effect on the environment and society. As far as social and environmental concerns are concerned, the financial consequences of poor maintenance of the energy use and heat-generating IT systems turns the attention of stakeholders toward green computing (GC).

The term green computing means that computers/mobiles and their resources are designed and manufactured in such a way that they are environmentally sound and eco-friendly. More broadly, the researchers have defined GC as developing, designing, disposing, and using computing devices in such a way that their environmental effect is reduced. Hence, GC increases the lifetime of computing devices and reduces the carbo effects on the ecosystem.

In this chapter, we will discuss about three different GCCs which will contribute to green communication (G Comm.). The three different GCC are as follows:

i. Universal Asynchronous Receiver Transmitter (UART)
ii. Finite Impulse Response (FIR) Filter
iii. Packet Counter

3.7 UART

UART is the acronym for Universal Asynchronous Receiver Transmitter. It is a computer-dedicated hardware device, an IC, or a circuit which is embedded on any microcontroller. It is associated for asynchronous serial communication of data bits. It is not a communication protocol in contrast to I2C and SPI. Both I2C and SPI are communication protocols. UART was one of the initial communication devices, attached with teletypewriters for communication purpose. It was also regarded a primary hardware structure for the Internet service. The block diagram of UART is shown in Figure 3.1, which mainly has shift register, transmitting register, receiving register, and clock signal [9–11].

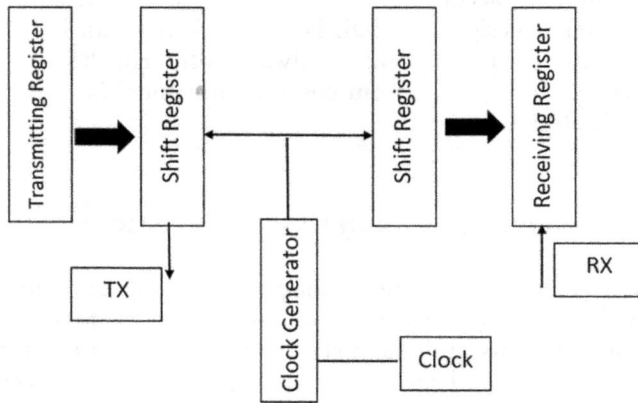

Figure 3.1 Block diagram of UART [12].

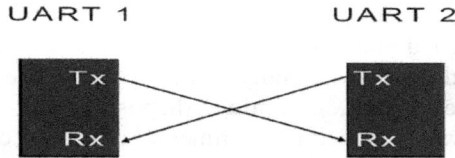

Figure 3.2 UART communication [13].

3.8 UART COMMUNICATION

The communication directly happens between two UARTs. In the communication at transmitter side, the UART converts parallel data bits to serial data bits, while at the receiving side, it does the opposite, i.e., serial to parallel conversion of data. The communication directly happens between Tx pin and Rx pin of the UART, as shown in Figure 3.2. For this communication, it requires only two wires. The UART is regarded as universal because its constraints like data speed, transfer speed, etc. can be configured any time.

The transmission of data bits is done asynchronously, which means that there is no need of any clock pulse for synchronizing transmitting and receiving bits in UART communication. It adds the start bits to the transmitting data packet and stops bits to the receiving data packet which is being communicated. If the receiving UART detects a start bit, the input bits begin to be read at a certain rate termed as the baud rate. The baud rate is the speed of transferring data bits, measured in terms of bits per second (bps). Both UARTs must work at approximately identical baud rates. The baud rates between the two UARTs can only vary by approximately 10%.

Steps involved in transmitting data with UART are as follows:

i. Parallel data is received at transmitting UART from the data bus.
ii. Start bit, stop bit, and parity bit(s) are being added by the transmitting UART to the data frame [13–15].
iii. Serial communication of data takes place between transmitting and receiving UART.
iv. Start, stop, and parity bits are discarded at receiving side.
v. Serial data is converted to parallel data and then sent to data bus by the receiving UART [16].

3.8.1 Advantages and disadvantages of UART

Advantages

i. Requires only two wires.
ii. Clock pulse is not needed.
iii. Parity bits check the error.

Disadvantages

i. Data frame size is restricted.
ii. Speed is less as equated to parallel communication.
iii. Baud rate must be matched.

3.8.2 VHDL/Verilog coding for UART

VHDL code for transmitting UART

```
library ieee;
use ieee.std_logic_1164.all;
use ieee.numeric_std.all;

entity UART_TX is
  generic (
    g_CLKS_PER_BIT : integer := 115      -- Needs to be
set correctly
    );
  port (
    i_Clk       : in  std_logic;
    i_TX_DV     : in  std_logic;
    i_TX_Byte   : in  std_logic_vector(7 downto 0);
```

```
     o_TX_Active : out std_logic;
     o_TX_Serial : out std_logic;
     o_TX_Done   : out std_logic
     );
end UART_TX;


architecture RTL of UART_TX is

  type t_SM_Main is (s_Idle, s_TX_Start_Bit,
                     s_TX_Data_Bits,
                     s_TX_Stop_Bit, s_Cleanup);
  signal r_SM_Main : t_SM_Main := s_Idle;

  signal r_Clk_Count : integer range 0 to g_CLKS_PER_
BIT-1 := 0;
  signal r_Bit_Index : integer range 0 to 7 := 0;   -- 8
Bits Total
  signal r_TX_Data  : std_logic_vector(7 downto 0) :=
(others => '0');
  signal r_TX_Done  : std_logic := '0';

begin


  p_UART_TX : process (i_Clk)
  begin
    if rising_edge(i_Clk) then

      case r_SM_Main is

        when s_Idle =>
          o_TX_Active <= '0';
          o_TX_Serial <= '1';          -- Drive Line
High for Idle
          r_TX_Done   <= '0';
          r_Clk_Count <= 0;
          r_Bit_Index <= 0;

          if i_TX_DV = '1' then
            r_TX_Data <= i_TX_Byte;
            r_SM_Main <= s_TX_Start_Bit;
          else
            r_SM_Main <= s_Idle;
          end if;
```

```
        -- Send out Start Bit. Start bit = 0
      when s_TX_Start_Bit =>
        o_TX_Active <= '1';
        o_TX_Serial <= '0';

        -- Wait g_CLKS_PER_BIT-1 clock cycles for
start bit to finish
        if r_Clk_Count < g_CLKS_PER_BIT-1 then
          r_Clk_Count <= r_Clk_Count + 1;
          r_SM_Main   <= s_TX_Start_Bit;
        else
          r_Clk_Count <= 0;
          r_SM_Main   <= s_TX_Data_Bits;
        end if;


        -- Wait g_CLKS_PER_BIT-1 clock cycles for data
bits to finish
      when s_TX_Data_Bits =>
        o_TX_Serial <= r_TX_Data(r_Bit_Index);

        if r_Clk_Count < g_CLKS_PER_BIT-1 then
          r_Clk_Count <= r_Clk_Count + 1;
          r_SM_Main   <= s_TX_Data_Bits;
        else
          r_Clk_Count <= 0;

          -- Check if we have sent out all bits
          if r_Bit_Index < 7 then
            r_Bit_Index <= r_Bit_Index + 1;
            r_SM_Main   <= s_TX_Data_Bits;
          else
            r_Bit_Index <= 0;
            r_SM_Main   <= s_TX_Stop_Bit;
          end if;
        end if;


        -- Send out Stop bit.  Stop bit = 1
      when s_TX_Stop_Bit =>
        o_TX_Serial <= '1';

        -- Wait g_CLKS_PER_BIT-1 clock cycles for
Stop bit to finish
        if r_Clk_Count < g_CLKS_PER_BIT-1 then
          r_Clk_Count <= r_Clk_Count + 1;
          r_SM_Main   <= s_TX_Stop_Bit;
```

```
        else
          r_TX_Done   <= '1';
          r_Clk_Count <= 0;
          r_SM_Main   <= s_Cleanup;
        end if;


      -- Stay here 1 clock
      when s_Cleanup =>
        o_TX_Active <= '0';
        r_TX_Done   <= '1';
        r_SM_Main   <= s_Idle;


      when others =>
        r_SM_Main <= s_Idle;

    end case;
  end if;
 end process p_UART_TX;

 o_TX_Done <= r_TX_Done;

end RTL;
```

VHDL code for receiving UART

```
library ieee;
use ieee.std_logic_1164.ALL;
use ieee.numeric_std.all;

entity UART_RX is
  generic (
    g_CLKS_PER_BIT : integer := 115      -- Needs to be
set correctly
    );
  port (
    i_Clk        : in  std_logic;
    i_RX_Serial  : in  std_logic;
    o_RX_DV      : out std_logic;
    o_RX_Byte    : out std_logic_vector(7 downto 0)
    );
end UART_RX;

architecture rtl of UART_RX is
```

```
   type t_SM_Main is (s_Idle, s_RX_Start_Bit,
                      s_RX_Data_Bits,
                      s_RX_Stop_Bit, s_Cleanup);
   signal r_SM_Main : t_SM_Main := s_Idle;

   signal r_RX_Data_R : std_logic := '0';
   signal r_RX_Data   : std_logic := '0';

   signal r_Clk_Count : integer range 0 to g_CLKS_PER_
BIT-1 := 0;
   signal r_Bit_Index : integer range 0 to 7 := 0;   -- 8
Bits Total
   signal r_RX_Byte   : std_logic_vector(7 downto 0) :=
(others => '0');
   signal r_RX_DV     : std_logic := '0';

begin

   -- Purpose: Double-register the incoming data.
   -- This allows it to be used in the UART RX Clock
Domain.
   -- (It removes problems caused by metastabiliy)
   p_SAMPLE : process (i_Clk)
   begin
     if rising_edge(i_Clk) then
       r_RX_Data_R <= i_RX_Serial;
       r_RX_Data   <= r_RX_Data_R;
     end if;
   end process p_SAMPLE;


   -- Purpose: Control RX state machine
   p_UART_RX : process (i_Clk)
   begin
     if rising_edge(i_Clk) then

       case r_SM_Main is

         when s_Idle =>
           r_RX_DV     <= '0';
           r_Clk_Count <= 0;
           r_Bit_Index <= 0;

           if r_RX_Data = '0' then        -- Start bit
detected
             r_SM_Main <= s_RX_Start_Bit;
           else
```

```
         r_SM_Main  <= s_Idle;
       end if;


       -- Check middle of start bit to make sure it's
still low
       when s_RX_Start_Bit =>
         if r_Clk_Count = (g_CLKS_PER_BIT-1)/2 then
           if r_RX_Data = '0' then
             r_Clk_Count <= 0;  -- reset counter since
we found the middle
             r_SM_Main   <= s_RX_Data_Bits;
           else
             r_SM_Main   <= s_Idle;
           end if;
         else
           r_Clk_Count <= r_Clk_Count + 1;
           r_SM_Main   <= s_RX_Start_Bit;
         end if;


       -- Wait g_CLKS_PER_BIT-1 clock cycles to sample
serial data
       when s_RX_Data_Bits =>
         if r_Clk_Count < g_CLKS_PER_BIT-1 then
           r_Clk_Count <= r_Clk_Count + 1;
           r_SM_Main   <= s_RX_Data_Bits;
         else
           r_Clk_Count             <= 0;
           r_RX_Byte(r_Bit_Index) <= r_RX_Data;

           -- Check if we have sent out all bits
           if r_Bit_Index < 7 then
             r_Bit_Index <= r_Bit_Index + 1;
             r_SM_Main   <= s_RX_Data_Bits;
           else
             r_Bit_Index <= 0;
             r_SM_Main   <= s_RX_Stop_Bit;
           end if;
         end if;


       -- Receive Stop bit.  Stop bit = 1
       when s_RX_Stop_Bit =>
         -- Wait g_CLKS_PER_BIT-1 clock cycles for
Stop bit to finish
         if r_Clk_Count < g_CLKS_PER_BIT-1 then
```

```
                r_Clk_Count  <= r_Clk_Count + 1;
                r_SM_Main    <= s_RX_Stop_Bit;
             else
                r_RX_DV      <= '1';
                r_Clk_Count <= 0;
                r_SM_Main    <= s_Cleanup;
             end if;


          -- Stay here 1 clock
          when s_Cleanup =>
            r_SM_Main <= s_Idle;
            r_RX_DV   <= '0';


          when others =>
            r_SM_Main <= s_Idle;

      end case;
    end if;
  end process p_UART_RX;

  o_RX_DV   <= r_RX_DV;
  o_RX_Byte <= r_RX_Byte;

end rtl;
```

VHDL code for UART

```
LIBRARY ieee;
USE ieee.std_logic_1164.all;

ENTITY uart IS
  GENERIC(
    clk_freq  :  INTEGER     := 50_000_000;  --frequency
of system clock in Hertz
    baud_rate :  INTEGER     := 19_200;      --data link
baud rate in bits/second
    os_rate   :  INTEGER     := 16;
--oversampling rate to find center of receive bits
(in samples per baud period)
    d_width   :  INTEGER     := 8;           --data bus
width
    parity    :  INTEGER     := 1;           --0 for no
parity, 1 for parity
```

```
      parity_eo : STD_LOGIC  := '0');           --'0' for
even, '1' for odd parity
  PORT(
    clk      : IN   STD_LOGIC;
--system clock
    reset_n  : IN   STD_LOGIC;
--ascynchronous reset
    tx_ena   : IN   STD_LOGIC;
--initiate transmission
    tx_data  : IN   STD_LOGIC_VECTOR(d_width-1
DOWNTO 0);  --data to transmit
    rx       : IN   STD_LOGIC;
--receive pin
    rx_busy  : OUT  STD_LOGIC;
--data reception in progress
    rx_error : OUT  STD_LOGIC;
--start, parity, or stop bit error detected
    rx_data  : OUT  STD_LOGIC_VECTOR(d_width-1
DOWNTO 0);  --data received
    tx_busy  : OUT  STD_LOGIC;
--transmission in progress
    tx       : OUT  STD_LOGIC);
--transmit pin
END uart;

ARCHITECTURE logic OF uart IS
  TYPE   tx_machine IS(idle, transmit);
--tranmit state machine data type
  TYPE   rx_machine IS(idle, receive);
--receive state machine data type
  SIGNAL tx_state     : tx_machine;
--transmit state machine
  SIGNAL rx_state     : rx_machine;
--receive state machine
  SIGNAL baud_pulse   : STD_LOGIC := '0';
--periodic pulse that occurs at the baud rate
  SIGNAL os_pulse     : STD_LOGIC := '0';
--periodic pulse that occurs at the oversampling rate
  SIGNAL parity_error : STD_LOGIC;
--receive parity error flag
  SIGNAL rx_parity    : STD_LOGIC_VECTOR(d_width
DOWNTO 0);  --calculation of receive parity
  SIGNAL tx_parity    : STD_LOGIC_VECTOR(d_width
DOWNTO 0);  --calculation of transmit parity
  SIGNAL rx_buffer    : STD_LOGIC_VECTOR(parity+d_
width DOWNTO 0) := (OTHERS => '0');   --values received
```

```
  SIGNAL tx_buffer      :   STD_LOGIC_VECTOR(parity+d_
width+1 DOWNTO 0) := (OTHERS => '1'); --values to be
transmitted
BEGIN

  --generate clock enable pulses at the baud rate and
the oversampling rate
  PROCESS(reset_n, clk)
    VARIABLE count_baud :   INTEGER RANGE 0 TO clk_freq/
baud_rate-1 := 0;         --counter to determine baud
rate period
    VARIABLE count_os   :   INTEGER RANGE 0 TO clk_freq/
baud_rate/os_rate-1 := 0; --counter to determine
oversampling period
  BEGIN
    IF(reset_n = '0') THEN
--asynchronous reset asserted
      baud_pulse <= '0';
--reset baud rate pulse
      os_pulse <= '0';
--reset oversampling rate pulse
      count_baud := 0;
--reset baud period counter
      count_os := 0;
--reset oversampling period counter
    ELSIF(clk'EVENT AND clk = '1') THEN
      --create baud enable pulse
      IF(count_baud < clk_freq/baud_rate-1) THEN
--baud period not reached
        count_baud := count_baud + 1;
--increment baud period counter
        baud_pulse <= '0';
--deassert baud rate pulse
      ELSE
--baud period reached
        count_baud := 0;
--reset baud period counter
        baud_pulse <= '1';
--assert baud rate pulse
        count_os := 0;
--reset oversampling period counter to avoid cumulative
error
      END IF;
      --create oversampling enable pulse
      IF(count_os < clk_freq/baud_rate/os_rate-1) THEN
--oversampling period not reached
```

```
        count_os := count_os + 1;
--increment oversampling period counter
        os_pulse <= '0';
--deassert oversampling rate pulse
      ELSE
--oversampling period reached
        count_os := 0;
--reset oversampling period counter
        os_pulse <= '1';
--assert oversampling pulse
      END IF;
    END IF;
  END PROCESS;


  --receive state machine
  PROCESS(reset_n, clk)
    VARIABLE rx_count :  INTEGER RANGE 0 TO parity+d_
width+2 := 0; --count the bits received
    VARIABLE os_count :  INTEGER RANGE 0 TO os_rate-1
:= 0;        --count the oversampling rate pulses
  BEGIN
    IF(reset_n = '0') THEN
--asynchronous reset asserted
      os_count := 0;
--clear oversampling pulse counter
      rx_count := 0;
--clear receive bit counter
      rx_busy <= '0';
--clear receive busy signal
      rx_error <= '0';
--clear receive errors
      rx_data <= (OTHERS => '0');
--clear received data output
      rx_state <= idle;
--put in idle state
    ELSIF(clk'EVENT AND clk = '1' AND os_pulse = '1')
THEN --enable clock at oversampling rate
      CASE rx_state IS
        WHEN idle =>
--idle state
          rx_busy <= '0';
--clear receive busy flag
          IF(rx = '0') THEN
--start bit might be present
            IF(os_count < os_rate/2) THEN
--oversampling pulse counter is not at start bit center
              os_count := os_count + 1;
--increment oversampling pulse counter
```

```
              rx_state <= idle;
--remain in idle state
          ELSE
--oversampling pulse counter is at bit center
              os_count := 0;
--clear oversampling pulse counter
              rx_count := 0;
--clear the bits received counter
              rx_busy <= '1';
--assert busy flag
              rx_state <= receive;
--advance to receive state
          END IF;
       ELSE
--start bit not present
          os_count := 0;
--clear oversampling pulse counter
          rx_state <= idle;
--remain in idle state
       END IF;
     WHEN receive =>
--receive state
       IF(os_count < os_rate-1) THEN
--not center of bit
          os_count := os_count + 1;
--increment oversampling pulse counter
          rx_state <= receive;
--remain in receive state
       ELSIF(rx_count < parity+d_width) THEN
--center of bit and not all bits received
          os_count := 0;
--reset oversampling pulse counter
          rx_count := rx_count + 1;
--increment number of bits received counter
          rx_buffer <= rx & rx_buffer(parity+d_width
DOWNTO 1);   --shift new received bit into receive
buffer
          rx_state <= receive;
--remain in receive state
       ELSE
--center of stop bit
          rx_data <= rx_buffer(d_width DOWNTO 1);
--output data received to user logic
          rx_error <= rx_buffer(0) OR parity_error OR
NOT rx;    --output start, parity, and stop bit error
flag
          rx_busy <= '0';
--deassert received busy flag
```

```
                rx_state <= idle;
--return to idle state
           END IF;
       END CASE;
     END IF;
  END PROCESS;

  --receive parity calculation logic
  rx_parity(0) <= parity_eo;
  rx_parity_logic: FOR i IN 0 to d_width-1 GENERATE
    rx_parity(i+1) <= rx_parity(i) XOR rx_buffer(i+1);
  END GENERATE;
  WITH parity SELECT  --compare calculated parity bit
with received parity bit to determine error
     parity_error <= rx_parity(d_width) XOR rx_
buffer(parity+d_width) WHEN 1,  --using parity
                     '0' WHEN OTHERS;
--not using parity

  --transmit state machine
  PROCESS(reset_n, clk)
    VARIABLE tx_count :  INTEGER RANGE 0 TO parity+d_
width+3 := 0;  --count bits transmitted
  BEGIN
    IF(reset_n = '0') THEN
--asynchronous reset asserted
       tx_count := 0;
--clear transmit bit counter
       tx <= '1';
--set tx pin to idle value of high
       tx_busy <= '1';
--set transmit busy signal to indicate unavailable
       tx_state <= idle;
--set tx state machine to ready state
    ELSIF(clk'EVENT AND clk = '1') THEN
       CASE tx_state IS
         WHEN idle =>
--idle state
           IF(tx_ena = '1') THEN
--new transaction latched in
             tx_buffer(d_width+1 DOWNTO 0) <=  tx_data &
'0' & '1';   --latch in data for transmission and
start/stop bits
             IF(parity = 1) THEN
--if parity is used
               tx_buffer(parity+d_width+1) <= tx_
parity(d_width);       --latch in parity bit from
parity logic
```

```
                    END IF;
                    tx_busy <= '1';
--assert transmit busy flag
                    tx_count := 0;
--clear transmit bit count
                    tx_state <= transmit;
--proceed to transmit state
                ELSE
--no new transaction initiated
                    tx_busy <= '0';
--clear transmit busy flag
                    tx_state <= idle;
--remain in idle state
                END IF;
            WHEN transmit =>
--transmit state
                IF(baud_pulse = '1') THEN
--beginning of bit
                    tx_count := tx_count + 1;
--increment transmit bit counter
                    tx_buffer <= '1' & tx_buffer(parity+d_
width+1 DOWNTO 1);   --shift transmit buffer to output
next bit
                END IF;
                IF(tx_count < parity+d_width+3) THEN
--not all bits transmitted
                    tx_state <= transmit;
--remain in transmit state
                ELSE
--all bits transmitted
                    tx_state <= idle;
--return to idle state
                END IF;
        END CASE;
        tx <= tx_buffer(0);
--output last bit in transmit transaction buffer
    END IF;
  END PROCESS;

  --transmit parity calculation logic
  tx_parity(0) <= parity_eo;
  tx_parity_logic: FOR i IN 0 to d_width-1 GENERATE
    tx_parity(i+1) <= tx_parity(i) XOR tx_data(i);
  END GENERATE;

END logic;
```

Verilog code for transmitting UART

```verilog
module uart_tx
  #(parameter CLKS_PER_BIT)
  (
   input       i_Clock,
   input       i_Tx_DV,
   input [7:0] i_Tx_Byte,
   output      o_Tx_Active,
   output reg  o_Tx_Serial,
   output      o_Tx_Done
   );

  parameter s_IDLE         = 3'b000;
  parameter s_TX_START_BIT = 3'b001;
  parameter s_TX_DATA_BITS = 3'b010;
  parameter s_TX_STOP_BIT  = 3'b011;
  parameter s_CLEANUP      = 3'b100;

  reg [2:0]   r_SM_Main     = 0;
  reg [7:0]   r_Clock_Count = 0;
  reg [2:0]   r_Bit_Index   = 0;
  reg [7:0]   r_Tx_Data     = 0;
  reg         r_Tx_Done     = 0;
  reg         r_Tx_Active   = 0;

  always @(posedge i_Clock)
    begin

      case (r_SM_Main)
        s_IDLE :
          begin
            o_Tx_Serial   <= 1'b1;        // Drive
Line High for Idle
            r_Tx_Done     <= 1'b0;
            r_Clock_Count <= 0;
            r_Bit_Index   <= 0;

            if (i_Tx_DV == 1'b1)
              begin
                r_Tx_Active <= 1'b1;
                r_Tx_Data   <= i_Tx_Byte;
                r_SM_Main   <= s_TX_START_BIT;
              end
            else
              r_SM_Main <= s_IDLE;
          end // case: s_IDLE
```

```
        // Send out Start Bit. Start bit = 0
        s_TX_START_BIT :
          begin
            o_Tx_Serial <= 1'b0;

            // Wait CLKS_PER_BIT-1 clock cycles for
start bit to finish
            if (r_Clock_Count < CLKS_PER_BIT-1)
              begin
                r_Clock_Count <= r_Clock_Count + 1;
                r_SM_Main     <= s_TX_START_BIT;
              end
            else
              begin
                r_Clock_Count <= 0;
                r_SM_Main     <= s_TX_DATA_BITS;
              end
          end // case: s_TX_START_BIT


        // Wait CLKS_PER_BIT-1 clock cycles for data
bits to finish
        s_TX_DATA_BITS :
          begin
            o_Tx_Serial <= r_Tx_Data[r_Bit_Index];

            if (r_Clock_Count < CLKS_PER_BIT-1)
              begin
                r_Clock_Count <= r_Clock_Count + 1;
                r_SM_Main     <= s_TX_DATA_BITS;
              end
            else
              begin
                r_Clock_Count <= 0;

                // Check if we have sent out all bits
                if (r_Bit_Index < 7)
                  begin
                    r_Bit_Index <= r_Bit_Index + 1;
                    r_SM_Main   <= s_TX_DATA_BITS;
                  end
                else
                  begin
                    r_Bit_Index <= 0;
                    r_SM_Main   <= s_TX_STOP_BIT;
```

```
                  end
               end
           end // case: s_TX_DATA_BITS


           // Send out Stop bit.  Stop bit = 1
           s_TX_STOP_BIT :
             begin
               o_Tx_Serial <= 1'b1;

               // Wait CLKS_PER_BIT-1 clock cycles for
Stop bit to finish
               if (r_Clock_Count < CLKS_PER_BIT-1)
                 begin
                   r_Clock_Count <= r_Clock_Count + 1;
                   r_SM_Main     <= s_TX_STOP_BIT;
                 end
               else
                 begin
                   r_Tx_Done     <= 1'b1;
                   r_Clock_Count <= 0;
                   r_SM_Main     <= s_CLEANUP;
                   r_Tx_Active   <= 1'b0;
                 end
             end // case: s_Tx_STOP_BIT


           // Stay here 1 clock
           s_CLEANUP :
             begin
               r_Tx_Done <= 1'b1;
               r_SM_Main <= s_IDLE;
             end


           default :
             r_SM_Main <= s_IDLE;

       endcase
     end

  assign o_Tx_Active = r_Tx_Active;
  assign o_Tx_Done   = r_Tx_Done;

endmodule

Verilog code for receiving UART:
module uart_rx
```

```
#(parameter CLKS_PER_BIT)
(
 input          i_Clock,
 input          i_Rx_Serial,
 output         o_Rx_DV,
 output [7:0]   o_Rx_Byte
 );

 parameter s_IDLE         = 3'b000;
 parameter s_RX_START_BIT = 3'b001;
 parameter s_RX_DATA_BITS = 3'b010;
 parameter s_RX_STOP_BIT  = 3'b011;
 parameter s_CLEANUP      = 3'b100;


 reg            r_Rx_Data_R = 1'b1;
 reg            r_Rx_Data   = 1'b1;

 reg [7:0]      r_Clock_Count = 0;
 reg [2:0]      r_Bit_Index   = 0; //8 bits total
 reg [7:0]      r_Rx_Byte     = 0;
 reg            r_Rx_DV       = 0;
 reg [2:0]      r_SM_Main     = 0;

 // Purpose: Double-register the incoming data.
 // This allows it to be used in the UART RX Clock
Domain.
 // (It removes problems caused by metastability)
 always @(posedge i_Clock)
   begin
     r_Rx_Data_R <= i_Rx_Serial;
     r_Rx_Data   <= r_Rx_Data_R;
   end


 // Purpose: Control RX state machine
 always @(posedge i_Clock)
   begin

     case (r_SM_Main)
       s_IDLE :
         begin
           r_Rx_DV       <= 1'b0;
           r_Clock_Count <= 0;
           r_Bit_Index   <= 0;

           if (r_Rx_Data == 1'b0)          // Start
bit detected
             r_SM_Main <= s_RX_START_BIT;
```

```
                else
                  r_SM_Main <= s_IDLE;
            end

          // Check middle of start bit to make sure it's
still low
          s_RX_START_BIT :
            begin
              if (r_Clock_Count == (CLKS_PER_BIT-1)/2)
                begin
                  if (r_Rx_Data == 1'b0)
                    begin
                      r_Clock_Count <= 0;  // reset
counter, found the middle
                      r_SM_Main    <= s_RX_DATA_BITS;
                    end
                  else
                    r_SM_Main <= s_IDLE;
                end
              else
                begin
                  r_Clock_Count <= r_Clock_Count + 1;
                  r_SM_Main     <= s_RX_START_BIT;
                end
            end // case: s_RX_START_BIT


          // Wait CLKS_PER_BIT-1 clock cycles to sample
serial data
          s_RX_DATA_BITS :
            begin
              if (r_Clock_Count < CLKS_PER_BIT-1)
                begin
                  r_Clock_Count <= r_Clock_Count + 1;
                  r_SM_Main     <= s_RX_DATA_BITS;
                end
              else
                begin
                  r_Clock_Count            <= 0;
                  r_Rx_Byte[r_Bit_Index] <= r_Rx_Data;

                  // Check if we have received all bits
                  if (r_Bit_Index < 7)
                    begin
                      r_Bit_Index <= r_Bit_Index + 1;
                      r_SM_Main   <= s_RX_DATA_BITS;
                    end
```

```
                    else
                      begin
                        r_Bit_Index <= 0;
                        r_SM_Main   <= s_RX_STOP_BIT;
                      end
                  end
              end // case: s_RX_DATA_BITS


            // Receive Stop bit.  Stop bit = 1
            s_RX_STOP_BIT :
              begin
                // Wait CLKS_PER_BIT-1 clock cycles for
Stop bit to finish
                if (r_Clock_Count < CLKS_PER_BIT-1)
                  begin
                    r_Clock_Count <= r_Clock_Count + 1;
                    r_SM_Main     <= s_RX_STOP_BIT;
                  end
                else
                  begin
                    r_Rx_DV        <= 1'b1;
                    r_Clock_Count <= 0;
                    r_SM_Main      <= s_CLEANUP;
                  end
              end // case: s_RX_STOP_BIT


            // Stay here 1 clock
            s_CLEANUP :
              begin
                r_SM_Main <= s_IDLE;
                r_Rx_DV   <= 1'b0;
              end


            default :
              r_SM_Main <= s_IDLE;

          endcase
        end

    assign o_Rx_DV   = r_Rx_DV;
    assign o_Rx_Byte = r_Rx_Byte;

  endmodule // uart_rx
```

Verilog code for UART

```
module uart
(
    parameter DATA_WIDTH = 8
)
(

    input wire                      clk,
    input wire                      rst,

    /*
     * AXI input
     */
    input wire [DATA_WIDTH-1:0]  s_axis_tdata,
    input wire                   s_axis_tvalid,
    output wire                   s_axis_tready,

    /*
     * AXI output
     */
    output wire [DATA_WIDTH-1:0]  m_axis_tdata,
    output wire                   m_axis_tvalid,
    input  wire                   m_axis_tready,

    /*
     * UART interface
     */
    input  wire                     rxd,
    output wire                     txd,

    /*
     * Status
     */
    output wire                     tx_busy,
    output wire                     rx_busy,
    output wire                     rx_overrun_error,
    output wire                     rx_frame_error,

    /*
     * Configuration
     */
    input  wire [15:0]              prescale

);

uart_tx #(
    .DATA_WIDTH(DATA_WIDTH)
)
```

```
uart_tx_inst (
    .clk(clk),
    .rst(rst),
    // axi input
    .s_axis_tdata(s_axis_tdata),
    .s_axis_tvalid(s_axis_tvalid),
    .s_axis_tready(s_axis_tready),
    // output
    .txd(txd),
    // status
    .busy(tx_busy),
    // configuration
    .prescale(prescale)
);

uart_rx #(
    .DATA_WIDTH(DATA_WIDTH)
)
uart_rx_inst (
    .clk(clk),
    .rst(rst),
    // axi output
    .m_axis_tdata(m_axis_tdata),
    .m_axis_tvalid(m_axis_tvalid),
    .m_axis_tready(m_axis_tready),
    // input
    .rxd(rxd),
    // status
    .busy(rx_busy),
    .overrun_error(rx_overrun_error),
    .frame_error(rx_frame_error),
    // configuration
    .prescale(prescale)
);

endmodule
```

3.9 FIR FILTER

FIR Filter can be coined up as finite impulse response (FIR) filter. It is one of the important filters used in digital world for DSP applications. As the name suggests, its impulse response is of finite time. Therefore, it can settle up to zero finite time. This filter is quite different from infinite impulse response (IIR) filter, which has an internal feedback loop. This filter rejects the DC components and permits the AC signals. A telephone line that serves as a filter is the finest example of the FIR filter. The frequency is reduced to a range considerably lower than the frequency of humans [17].

Table 3.2 Comparison between FIR and IIR filters

Property	FIR	IIR
Nature	Nonrecursive	Recursive
Efficiency	Less	More
Usage	Difficult	Easy
Feedback	Don't require feedback	Needs feedback
Stability	More stable	Less stable
Delay	Gives more delay	Delay is less
Sensitivity	Less sensitive	More sensitive
Memory	Consumes more memory	Consumes less memory
Controllability	Easy	Difficult

The method for designing a FIR filter is based on optimization of an ideal filter. The FIR filters reach the ideal characteristics by increasing the order. The designing begins with requirements of the FIR filter. The process used in the designing depends upon the execution and requirements. The design approaches have both benefits and drawbacks. It is therefore very important to choose the right FIR filter form. Since the FIR filter is efficient and simple, the window method is most widely used. The other approach is also very easy to use, but the stopband has a slight attenuation.

The comparison between the two filters is represented in Table 3.2.

3.10 ADVANTAGES OF FIR FILTERS

- It is constantly stable
- Simple
- Have linear phase response
- Easy to be optimized
- Non-causal
- Signal-to-noise ratio (SNR) is less
- FIR can be designed as both a recursive and nonrecursive filter
- Effective performance
- Robust

3.11 DISADVANTAGES

- Requires more memory
- Cannot be simulated as analog filter design
- Complex designing
- Difficult implementation

- Expensive
- Time delay is more

3.12 VHDL CODE FOR FIR FILTER

The VHDL code provided in this section will illustrate how to design a FIR filter on FPGA for G Comm.. We're going to use an example using a 32-tap FIR filter.

The design of FIR filter will be singled clocked. The fundamental of the design is rooted in the function sum of products (SOP) which is used for accumulating a series of multiplications. The present data input signal is combined up with the data_table. The data inputs are sampled over the last N clock cycles (in this example N is 32), with respect to the shift_fifo function. The coefficients of the filter reach on the two-dimensional b input which are stored in the coefficient_table_var array. For the SOP function, the data set of the data_table has to be reversed. The VHDL code for the FIR filter is presented below:

```
library ieee;
  use ieee.std_logic_1164.all;
  use ieee.numeric_std.all;
use work.types.all;

library maths;
  use maths.maths_class.all;
library matrix;
  use matrix.matrix_class.all;

entity FIR_32tap_8_8 is
port (
  a : in   std_logic_vector(7 downto 0);
  b : in   logic_8_vector(31 downto 0);
  clock : in   std_logic;
  reset : in   std_logic;
  y : out std_logic_vector(20 downto 0)
);
end FIR_32tap_8_8;

architecture behavioural of FIR_32tap_8_8 is

  constant number_of_taps: integer := 32;

  signal data_table: single_vector(number_of_taps-1
downto 0);
```

```
   signal coefficient_table: single_vector(number_of_
taps-1 downto 0);

begin

   -- y <= sum_over (0, k-1, a((k-1)-i), b(i))
   -- coefficient_table <= b;

   fir_algorithm: process (clock)
     variable data_out : single;
     variable fir_result : single;
     variable data_table_var: single_vector(number_of_
taps-1 downto 0);
     -- the coeff table assignment really ought to be
handled at the entity interface
     variable coefficient_table_var: single_
vector(number_of_taps-1 downto 0);
     variable tmp : single_vector(number_of_taps-1
downto 0);
     variable tmp2 : single;
     variable tmp3 : single_vector(number_of_taps-1
downto 0);
     variable tmp4 : integer;
     variable num_taps_minus_1 : integer;
     variable y_result : signed(20 downto 0);
   begin
     if rising_edge(clock) then
       -- data_table_var := data_table(number_of_taps-1)
& data_table(number_of_taps-2 downto 0);
       -- putting the coeff table in a loop like this
allows dynamic coeff updating
       for i in 0 to number_of_taps-1 loop
         coefficient_table_var(i) :=
single(to_integer(unsigned(b(i))))/127.0;
       end loop;
       -- tmp := reverse_order(data_table_var);

       -- tmp2 := 0.15;  + to_integer(a);
       data_table_var := data_table;
       tmp2 := single(to_integer(signed(a)));
       data_table_var := shift_fifo (data_table_var,
tmp2); -- fifo =>  data_in =>
       data_table <= data_table_var;

       -- tmp3 := reverse_order(data_table_var);
       -- tmp4 := 0;
       num_taps_minus_1 := number_of_taps-1;
       fir_result := sum_of_products (
```

```
         lower_limit => 0,
         upper_limit => number_of_taps-1,
         a_in => reverse_order(data_table_var),
         b_in => coefficient_table_var
      );
      y_result := to_signed(integer(fir_result),
y_result'length);
      y <= std_logic_vector(y_result);
   end if;
  end process;
end behavioural;
```

Verilog code

```verilog
module FIR (
    input clk,
    input reset,
    input signed [15:0] s_axis_fir_tdata,
    input [3:0] s_axis_fir_tkeep,
    input s_axis_fir_tlast,
    input s_axis_fir_tvalid,
    input m_axis_fir_tready,
    output reg m_axis_fir_tvalid,
    output reg s_axis_fir_tready,
    output reg m_axis_fir_tlast,
    output reg [3:0] m_axis_fir_tkeep,
    output reg signed [31:0] m_axis_fir_tdata
    );

    always @ (posedge clk)
        begin
            m_axis_fir_tkeep <= 4'hf;
        end
    always @ (posedge clk)
        begin
            if (s_axis_fir_tlast == 1'b1)
                begin
                    m_axis_fir_tlast <= 1'b1;
                end
            else
                begin
                    m_axis_fir_tlast <= 1'b0;
                end
        end
```

```
    // 15-tap FIR
    reg enable_fir, enable_buff;
    reg [3:0] buff_cnt;
    reg signed [15:0] in_sample;
    reg signed [15:0] buff0, buff1, buff2, buff3,
buff4, buff5, buff6, buff7, buff8, buff9, buff10,
buff11, buff12, buff13, buff14;
    wire signed [15:0] tap0, tap1, tap2, tap3, tap4,
tap5, tap6, tap7, tap8, tap9, tap10, tap11, tap12,
tap13, tap14;
    reg signed [31:0] acc0, acc1, acc2, acc3, acc4,
acc5, acc6, acc7, acc8, acc9, acc10, acc11, acc12,
acc13, acc14;


    /* Taps for LPF running @ 1MSps with a cutoff freq
of 400kHz*/
    assign tap0 = 16'hFC9C;  // twos(-0.0265 * 32768) =
0xFC9C
    assign tap1 = 16'h0000;  // 0
    assign tap2 = 16'h05A5;  // 0.0441 * 32768 =
1445.0688 = 1445 = 0x05A5
    assign tap3 = 16'h0000;  // 0
    assign tap4 = 16'hF40C;  // twos(-0.0934 * 32768) =
0xF40C
    assign tap5 = 16'h0000;  // 0
    assign tap6 = 16'h282D;  // 0.3139 * 32768 =
10285.8752 = 10285 = 0x282D
    assign tap7 = 16'h4000;  // 0.5000 * 32768 = 16384
= 0x4000
    assign tap8 = 16'h282D;  // 0.3139 * 32768 =
10285.8752 = 10285 = 0x282D
    assign tap9 = 16'h0000;  // 0
    assign tap10 = 16'hF40C; // twos(-0.0934 * 32768) =
0xF40C
    assign tap11 = 16'h0000; // 0
    assign tap12 = 16'h05A5; // 0.0441 * 32768 =
1445.0688 = 1445 = 0x05A5
    assign tap13 = 16'h0000; // 0
    assign tap14 = 16'hFC9C; // twos(-0.0265 * 32768) =
0xFC9C

    /* This loop sets the tvalid flag on the output of
the FIR high once
     * the circular buffer has been filled with input
samples for the
     * first time after a reset condition. */
    always @ (posedge clk or negedge reset)
```

```
        begin
            if (reset == 1'b0) //if (reset == 1'b0 ||
tvalid_in == 1'b0)
                begin
                    buff_cnt <= 4'd0;
                    enable_fir <= 1'b0;
                    in_sample <= 8'd0;
                end
            else if (m_axis_fir_tready == 1'b0 ||
s_axis_fir_tvalid == 1'b0)
                begin
                    enable_fir <= 1'b0;
                    buff_cnt <= 4'd15;
                    in_sample <= in_sample;
                end
            else if (buff_cnt == 4'd15)
                begin
                    buff_cnt <= 4'd0;
                    enable_fir <= 1'b1;
                    in_sample <= s_axis_fir_tdata;
                end
            else
                begin
                    buff_cnt <= buff_cnt + 1;
                    in_sample <= s_axis_fir_tdata;
                end
        end

    always @ (posedge clk)
        begin
            if(reset == 1'b0 || m_axis_fir_tready ==
1'b0 || s_axis_fir_tvalid == 1'b0)
                begin
                    s_axis_fir_tready <= 1'b0;
                    m_axis_fir_tvalid <= 1'b0;
                    enable_buff <= 1'b0;
                end
            else
                begin
                    s_axis_fir_tready <= 1'b1;
                    m_axis_fir_tvalid <= 1'b1;
                    enable_buff <= 1'b1;
                end
        end

    /* Circular buffer bring in a serial input sample
stream that
```

```
        * creates an array of 15 input samples for the 15
taps of the filter. */
    always @ (posedge clk)
        begin
            if(enable_buff == 1'b1)
                begin
                    buff0 <= in_sample;
                    buff1 <= buff0;
                    buff2 <= buff1;
                    buff3 <= buff2;
                    buff4 <= buff3;
                    buff5 <= buff4;
                    buff6 <= buff5;
                    buff7 <= buff6;
                    buff8 <= buff7;
                    buff9 <= buff8;
                    buff10 <= buff9;
                    buff11 <= buff10;
                    buff12 <= buff11;
                    buff13 <= buff12;
                    buff14 <= buff13;
                end
            else
                begin
                    buff0 <= buff0;
                    buff1 <= buff1;
                    buff2 <= buff2;
                    buff3 <= buff3;
                    buff4 <= buff4;
                    buff5 <= buff5;
                    buff6 <= buff6;
                    buff7 <= buff7;
                    buff8 <= buff8;
                    buff9 <= buff9;
                    buff10 <= buff10;
                    buff11 <= buff11;
                    buff12 <= buff12;
                    buff13 <= buff13;
                    buff14 <= buff14;
                end
        end

    /* Multiply stage of FIR */
    always @ (posedge clk)
        begin
            if (enable_fir == 1'b1)
                begin
```

```
                    acc0 <= tap0 * buff0;
                    acc1 <= tap1 * buff1;
                    acc2 <= tap2 * buff2;
                    acc3 <= tap3 * buff3;
                    acc4 <= tap4 * buff4;
                    acc5 <= tap5 * buff5;
                    acc6 <= tap6 * buff6;
                    acc7 <= tap7 * buff7;
                    acc8 <= tap8 * buff8;
                    acc9 <= tap9 * buff9;
                    acc10 <= tap10 * buff10;
                    acc11 <= tap11 * buff11;
                    acc12 <= tap12 * buff12;
                    acc13 <= tap13 * buff13;
                    acc14 <= tap14 * buff14;
                end
            end

    /* Accumulate stage of FIR */
    always @ (posedge clk)
        begin
            if (enable_fir == 1'b1)
                begin
                    m_axis_fir_tdata <= acc0 + acc1 +
acc2 + acc3 + acc4 + acc5 + acc6 + acc7 + acc8 + acc9 +
acc10 + acc11 + acc12 + acc13 + acc14;
                end
        end
endmodule
```

3.13 PACKET COUNTER

It defines the size of data transmission and the approximate charges of data communications with PacketWIN in real time. It is done by installing the packet counter program in the PC. It is one of the important tools used for data communication within PC and mobile devices [18]. There are several features of packet counter which are listed below:

a. Communications charge: The data communications size and the communications charge according to the selected rate can be presented in real time or after the communication, respectively.

b. Rate configuration: The customer's contracted rate can be entered in advance, and communications charges can be presented based on this rate.

c. Data size monitoring: A warning window can be presented when the data size exceeds the predetermined number of bytes (number of packets).
d. The communications log: One can save a conversation log in CSV format (start time, end time, communications volume, communications charges, and communications destination address).

Verilog code for packet counter

```
module packetcounter(packetin, clock, count,
packetout);
    input [31:0] packetin;
        input clock;
    output [5:0] count;
    output [31:0]  packetout;

reg [5:0] count;
reg [31:0]  packetout;

always@(posedge clock)
begin
count<=count+1;
packetout<=packetin;
end

endmodule
```

REFERENCES

1. Thomas, Donald, and Philip Moorby. *The verilog® hardware description language.* Springer Science & Business Media, Berlin, 2008.
2. Shahdad, Moe, Roger Lipsett, Erich Marschner, Kellye Sheehan, and Howard Cohen. "VHSIC hardware description language." *Computer* 18, no. 2 (1985): 94–103.
3. Harris, Sarah L. and David Harris, "Hardware description languages: An overview. ScienceDirect Topics. https://www.sciencedirect.com/topics/computer-science/hardware-description-languages. Accessed on 30 June 2023.
4. What is a hardware description language (HDL)? Technical articles. All About Circuits. https://www.allaboutcircuits.com/technical-articles/what-is-a-hardware-description-language-hdl/. Accessed on 30 June 2023.
5. Hardware description language | VLSI tutorial. Mepits. https://www.mepits.com/tutorial/143/vlsi/hardware-description-language. Accessed on 30 June 2023.
6. Mantooth, H. Alan, and Mike Fiegenbaum. *Modeling with an analog hardware description language.* Kluwer Academic, Dordrecht, 1994.

7. Galloway, David. "The transmogrifier C hardware description language and compiler for FPGAs." In *Proceedings IEEE Symposium on FPGAs for Custom Computing Machines,* Swansea, pp. 136–144. IEEE, 1995.

8. Li, Yanbing, and Miriam Leeser. "HML, a novel hardware description language and its translation to VHDL." *IEEE Transactions on Very Large Scale Integration (VLSI) Systems* 8, no. 1 (2000): 1–8.

9. Kumar, Keshav, Amanpreet Kaur, S. N. Panda, and Bishwajeet Pandey. "Effect of different nano meter technology-based FPGA on energy efficient UART design." In *2018 8th International Conference on Communication Systems and Network Technologies (CSNT),* India, pp. 1–4. IEEE, 2018.

10. Kumar, Keshav, Amanpreet Kaur, Bishwajeet Pandey, and S. N. Panda. "Low power UART design using different nanometer technology based FPGA." In *2018 8th International Conference on Communication Systems and Network Technologies (CSNT),* India, pp. 1–3. IEEE, 2018.

11. Kumar, Keshav, Bishwajeet Pandey, Amit Kant Pandit, Y. A. Baker El-Ebiary, Salameh A. Mjlae, and Samer Bamansoor. "Design of low power transceiver on Spartan-3 and Spartan-6 FPGA." *International Journal of Innovative Technology and Exploring Engineering* 8, no. 12S2 (2019): 27–30.

12. Interface - UDR (UART data register) issues. Stack Overflow. https://stackoverflow.com/questions/31904609/udr-uart-data-registerissues. Accessed on 30 June 2023.

13. Basics of UART explained - Communication protocol and its applications. ElProCus. https://www.elprocus.com/basics-of-uart-communication-block-diagram-applications/. Accessed on 30 June 2023.

14. UART (Universal asynchronous receiver transmitter) communication. https://electricalfundablog.com/uart-universal-asynchronous-receiver-transmitter-communication/. Accessed on 30 June 2023.

15. Basics of UART communication. Circuit Basics. https://www.circuitbasics.com/basics-uart-communication/. Accessed on 30 June 2023.

16. What is UART (Universal asynchronous receiver/transmitter)? Definition. https://www.techtarget.com/whatis/definition/UART-Universal-Asynchronous-Receiver-Transmitter. Accessed on 30 June 2023.

17. Pandey, Bishwajeet, Bhagwan Das, Amanpreet Kaur, Tanesh Kumar, Abdul Moid Khan, D. M. Akbar Hussain, and Geetam Singh Tomar. "Performance evaluation of FIR filter after implementation on different FPGA and SOC and its utilization in communication and network." *Wireless Personal Communications* 95, no. 2 (2017): 375–389.

18. Packet counter | Software download. au. https://www.au.com/english/mobile/service/mobile-communications/soft-download/packet-counter/. Accessed on 30 June 2023.

Chapter 4

LVCMOS-based UART for GCC

LIST OF ABBREVIATIONS

BUFG	Global Buffers
Clk	Clocks
DP	Dynamic Power
FF	Flip-Flops
FPGA	Field Programmable Gate Array
GC	Green Computing
GCC	Green Communication Computing
G. Comm.	Green Communication
GPS	Global Positioning System
IC	Integrated Circuit
IO	Input Output
JT	Junction Temperature
LP	Leakage Power
LUT	Look Up Tables
LVCMOS	Low-Voltage Complementary Metal-Oxide Semiconductor
RFID	Radio-Frequency Identification
S/G	Signal
SP	Static Power
TM	Thermal Margin
TP	Total Power
TPC	Total Power Consumption
UART	Universal Asynchronous Receiver Transmitter
ϑJA	Effective Thermal Resistance to Air

4.1 INTRODUCTION TO UART

UART is the acronym for universal asynchronous receiver transmitter. It is a predefined hardware used for serial communication of data bits. The UART device is generally integrated on the CPU of the computer or microcontroller, and sometimes it can be dedicated to an IC (integrated circuit) [1,2]. It is one of the most modest and informal techniques to perform the serial communication of data. The applications of UART can be seen in many industries like mobile communication, RFID communication, GPS, wireless communication, etc. The key purpose of integrating the UART hardware to microcontrollers is that serial communication involves only two wires for data transfer. In the transmitter unit of UART, parallel to serial conversion of data takes place, while at receiver unit, parallel to serial conversion of bits takes place [3]. Since it is hardware, it acts as a bridge between the communication protocols and microcontrollers, which can be seen in Figure 4.1.

The time constraint of UART must be matched for executing communication. Special bits are used in UART communication at the time of commencing and completion of each data packet. This process helps in synchronizing the transmitter and receiver units. The communication between two UARTs is shown in Figure 4.2.

The transmission in UART communication takes place in the form of packets, which is shown in Figure 4.3. These packets of data are categorized into three subsections as follows:

a. Start bits (1 or 2 bits)
b. Data bits (5–9 bits)
c. Parity bits (0 or 1 bits)
d. Stop bits (1 or 2 bits)

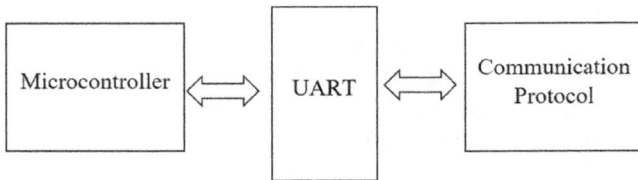

Figure 4.1 UART as a bridge between the communication protocols and microcontrollers.

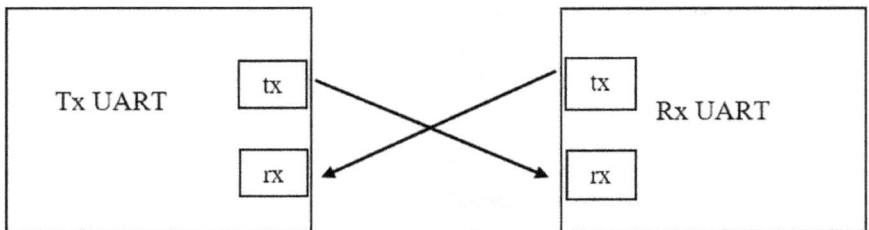

Figure 4.2 Communication of two UARTs.

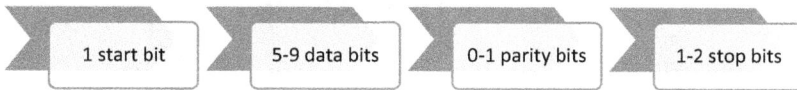

Figure 4.3 Structure of data packet of UART [4].

The different phases involved in UART communication are as follows:

a. Parallel data is received at transmission unit from the data bus.
b. Start, parity, and stop bits are added to the packets.
c. Serial transmission of packets takes place from transmitting unit to receiving unit.
d. At receiver side, start, parity, and stop bits are discarded.
e. At the receiving side, serial to parallel conversion of data takes place and hence data is transferred to data bus.

4.2 INTRODUCTION TO LVCMOS IO STANDARD

In FPGA case, the IO standards are used to match the impedance of input line to output line and input port to output port. Impedance matching is a technique that is used in electronics circuit world so that the impedance of the circuit is matched to minimize the power consumption. Here to minimize the power consumption of the UART circuit on Kintex-7 FPGA device, LVCMOS IO standard is used [5,6]. LVCMOS is the acronym for low-voltage complementary metal oxide. It is a low-voltage class of CMOS technology digital IC. There are five different types of LVCMOS IO standard as follows:

- LVCMOS 12: Here 12 means the operating voltage of this IO standard 1.2 V
- LVCMOS 15: Here 15 means the operating voltage of this IO standard 1.5 V
- LVCMOS 18: Here 18 means the operating voltage of this IO standard 1.8 V
- LVCMOS 12: Here 25 means the operating voltage of this IO standard 2.5 V
- LVCMOS 12: Here 33 means the operating voltage of this IO standard 3.3 V

4.3 INTEGRATION OF UART WITH FPGA

This section will elaborate on the implementation of UART on FPGA. Here the UART is implemented on Kintex-7 FPGA device. The register transfer logic (RTL) design is described in Figure 4.4.

In Figure 4.4, there are two UART: one is for transmitting the signal, and the other is for receiving the signal. At the input of transmitting end, there are five wires (clk, 16-bit perscale, rst, s_axis_tdata 8 bit, and s_axix_tvalid), while at output side, there are three wires (busy, txd, and s_axix_tready). Similarly, at the input of receiving side, there are six wires (clk, 16-bit perscale, rst, m_axis_tready, rst, and rxd), while at output side, there are five wires (busy, frame_error, m_axis_tdata 8 bit, m_axis_tvalid, and overrun_error) [7–9].

4.4 RESOURCE UTILIZATION

There are some FPGA resources used in the implementation of UART on Kintex-7 device. These resources are look up tables (LUTs), flip-flops (FF), input output (IO), and global buffers (BUFG). In the implementation of UART, Kintex-7 consumes 79 FF, 110 LUTs, 44 IO, and 1 BUFG. The resource utilization of Kintex-7 is described in Table 4.1, and the percentage of resource consumed is shown in Figure 4.5 [10–12].

4.5 THERMAL PROPERTIES

In the implementation of UART, there might be a chance of heating up of device; sometimes the devices underperform if there is mismatch in

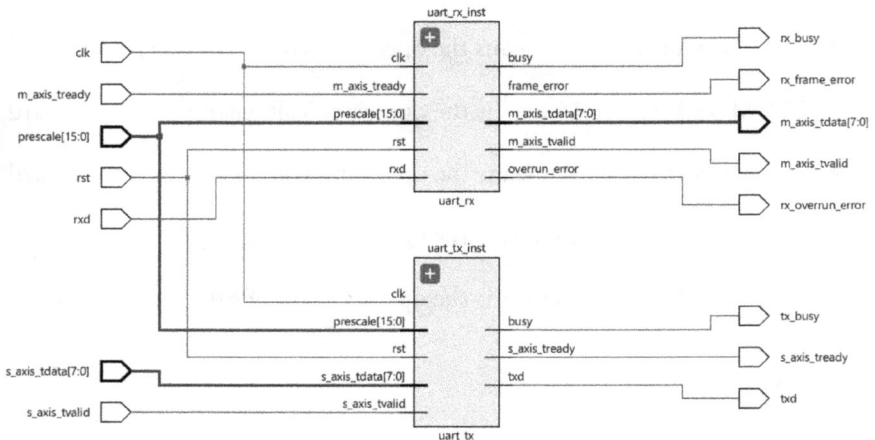

Figure 4.4 RTL design of UART.

Table 4.1 Resource utilization of UART on Kintex-7

Resources	Utilization	Available	Utilization %
FF	79	43,3920	0.02
IO	44	304	14.47
LUT	110	21,6960	0.05
BUFG	1	256	0.39

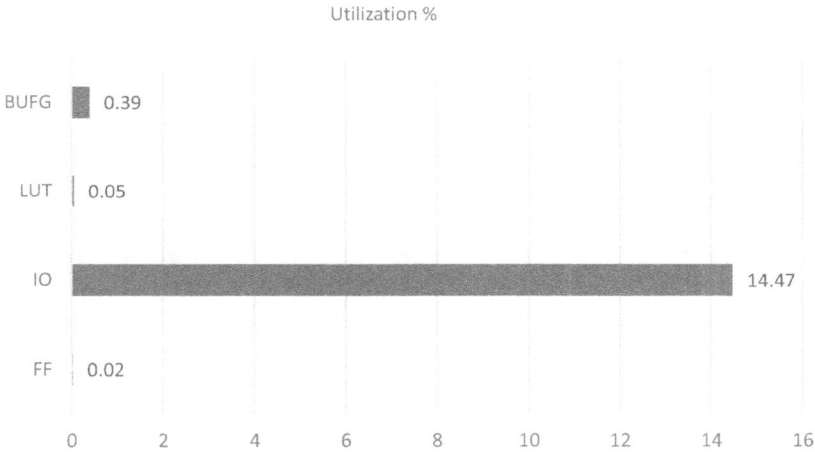

Utilization %

BUFG	0.39
LUT	0.05
IO	14.47
FF	0.02

0   2   4   6   8   10   12   14   16

Figure 4.5 Percentage of resource consumed.

Table 4.2 Thermal properties for Kintex-7 device
with LVCMOS IO standards

Thermal properties	LVCMOS IO standards
JT (°C)	26.1
TM (°C)	73.9
ϑJA (°C/W)	2.3

junction temperature, thermal margin. All such properties that make
the proper working of UART on Kintex-7 can be considered as thermal
properties. The thermal properties that are considered here are junction
temperature (JT), effective thermal resistance to air (ϑJA), and thermal
margin (TM) [13–15]. Here the thermal properties are tested for five different LVCMOS IO standards. Thermal properties are the same for all the
tested IO standards. The thermal properties with LVCMOS IO standards
required for Kintex-7 device in the implementation of UART are shown
in Table 4.2.

4.6 POWER ANALYSIS

The power consumption is one of the crucial issues which the whole electronics sector is facing. We all are trying to build such a system that consumes low power. Lower the power consumption, the better the life system of the device. Here we are calculating the power consumption of UART implementation on Kintex-7 device by matching its impedance with LVCMOS IO Standard. The total power consumption of the device is the sum of device dynamic and static power [16,17] as mathematically expressed in Equation 4.1.

$$TP = SP + DP \qquad (4.1)$$

where

TP = Total power
SP = Static Power
DP = Dynamic Power

The device SP comprises the sum up of Clocks (clk), IO, Logic, and Signal (S/G) power, and the DP is the leakage power (LP) of the device. The TP of the device is the same for LVCMOS 12 and LVCMOS 15 IO standards. Also, the TP is the same for LVCMOS 25 and LVCMOS 33. It is different for LVCMOS 18.

4.6.1 Power analysis for LVCMOS 12 and LVCMOS 15

When the LVCMOS 12 and LVCMOS 15 IO standards are used for matching the impedance of the device, the TP consumption of the device observed is 0.487 W. The device SP is 0.468 W which is 96% of TP, and DP is 0.019 W which is 4% of TP consumption. The power consumption for LVCMOS 12 and LVCMOS 15 is illustrated in Figure 4.6. The graphical representation of TP consumption is described in Figure 4.7.

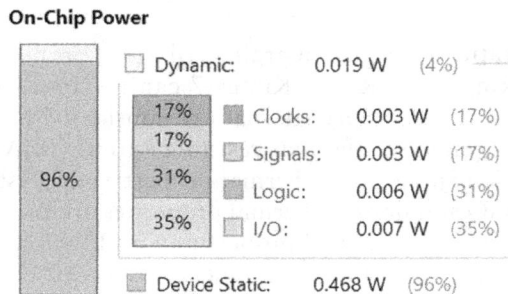

On-Chip Power

Dynamic:	0.019 W	(4%)
17% ■ Clocks:	0.003 W	(17%)
17% ■ Signals:	0.003 W	(17%)
31% ■ Logic:	0.006 W	(31%)
35% ■ I/O:	0.007 W	(35%)
96%		
■ Device Static:	0.468 W	(96%)

Figure 4.6 TP consumption for LVCMOS 12 and LVCMOS 15.

POWER CONSUMPTION FOR
LVCMOS 12 AND LVCMOS 15

IIII Power (W)

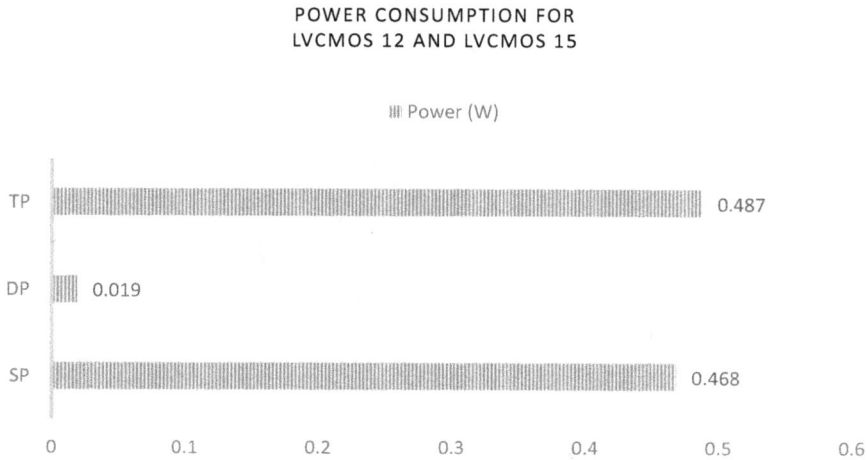

Figure 4.7 Graphical representation of TP consumption for LVCMOS I2 and LVCMOS I5.

4.6.2 Power analysis for **LVCMOS 18**

When the power analysis is done by matching the impedance with LVCMOS 18 IO standard, the SP observed is 0.468 W which is 96% of TP and the DP is 4% of TP which is 0.020 W. The TP is the summation of DP and SP which is 0.488 W as shown in Figure 4.8. The graphical representation of power consumption for LVCMOS 18 IO standard is shown in Figure 4.9.

4.6.3 Power analysis for **LVCMOS 25 and LVCMOS 33**

When the LVCMOS 25 and LVCMOS 33 IO standards are used for matching the impedance of the device, the TP consumption of the device observed is 0.484 W. The device SP is 0.468 W which is 97% of TP, and DP is 0.016 W which is 3% of TP consumption. The power consumption for LVCMOS 25 and LVCMOS 33 is illustrated in Figure 4.10. The graphical representation of TP consumption is described in Figure 4.11.

4.7 OBSERVATION AND COMPARATIVE ANALYSIS

From the power analysis section, it is observed that the SP for all the used IO standards remains same. There is only change in DP when there is change of IO standard. The TP consumption gets increased for LVCMOS 18 IO standard. But when the operating voltage of the IO standard gets increased, i.e., LVCMOS 25 and LVCMOS 33, the TP consumption decreases. The TP consumption for the used IO standards is shown in Table 4.3 and illustrated in Figure 4.12. From the power analysis, it is clearly observed that for getting low

On-Chip Power

☐ Dynamic:	0.020 W (4%)
17% ▣ Clocks:	0.003 W (17%)
17% ▣ Signals:	0.003 W (17%)
30% ▣ Logic:	0.006 W (30%)
36% ▢ I/O:	0.007 W (36%)
▣ Device Static:	0.468 W (96%)

96%

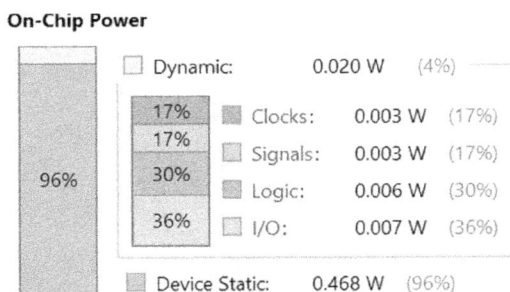

Figure 4.8 TP consumption for LVCMOS 18.

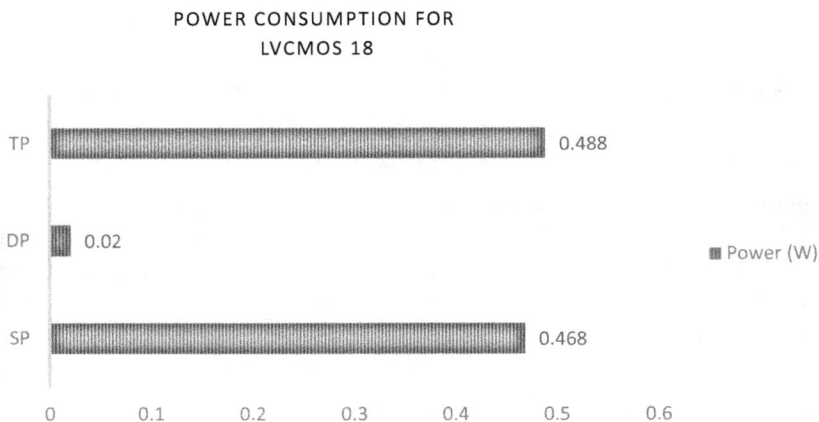

POWER CONSUMPTION FOR
LVCMOS 18

TP   0.488

DP   0.02

SP   0.468

■ Power (W)

0   0.1   0.2   0.3   0.4   0.5   0.6

Figure 4.9 Graphical representation of TP consumption for LVCMOS 18.

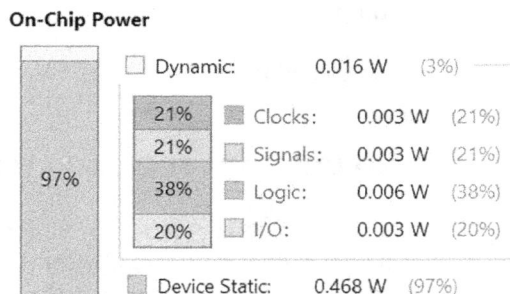

On-Chip Power

☐ Dynamic:	0.016 W (3%)
21% ▣ Clocks:	0.003 W (21%)
21% ▣ Signals:	0.003 W (21%)
38% ▣ Logic:	0.006 W (38%)
20% ▢ I/O:	0.003 W (20%)
▣ Device Static:	0.468 W (97%)

97%

Figure 4.10 TP consumption for LVCMOS 25 and LVCMOS 33.

POWER CONSUMPTION FOR
LVCMOS 25 AND LVCMOS 33

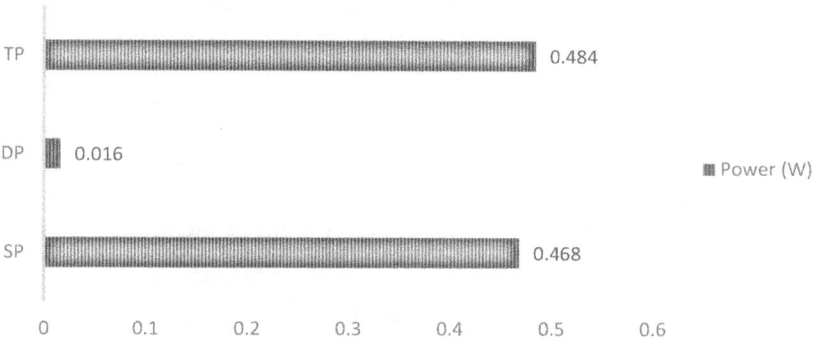

Figure 4.11 Graphical representation of TP consumption for LVCMOS 25 and LVCMOS 33.

Table 4.3 TP consumption for used IO standards

IO standards	TP (W)
LVCMOS 12	0.487
LVCMOS 15	0.487
LVCMOS 18	0.488
LVCMOS 25	0.484
LVCMOS 33	0.484

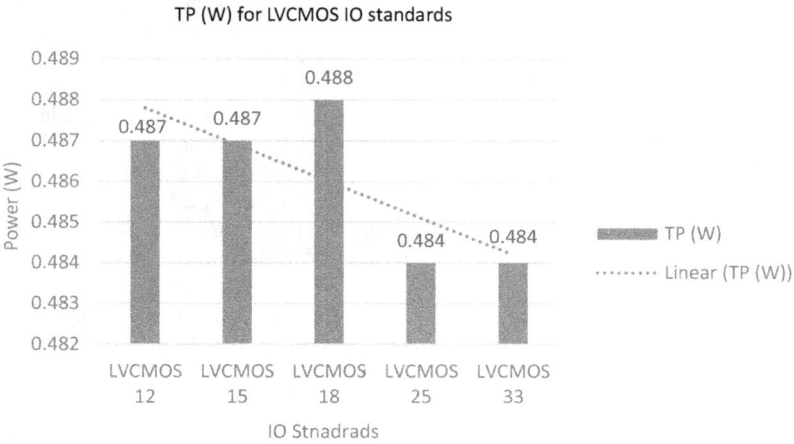

Figure 4.12 TP consumption for used IO standards.

power consumption of UART on Kintex-7 device, one should match the impedance of the device with LVCMOS 25 and LVCMOS 33 IO standards. For the other IO standards of LVCMOS family, the TP consumption is slightly higher as compared to LVCMOS 25 and LVCMOS 33 IO standards.

4.8 CONCLUSION

It has been observed that in past few years, the globe is facing huge problem of energy and power deficiency. The earth's natural resources may not live long. Someday it has to be vanished. In the context of such a huge problem, the idea of GC comes in people's mind. This chapter is a step in the area of GCC. In this chapter, a power-efficient model of UART is designed with the help of Kintex-7 device. To optimize the power utilization of UART, in this chapter the authors have used the impedance matching technique, for which they have used the LVCMOS IO standards. In FPGA, IO standards are used to match the input and output impedance of the circuit. In this chapter, the optimized power is observed for LVCMOS 25 and LVCMOS 33 IO standards as compared to the other LVCMOS IO standards. There is an increment in power consumption when the TP consumption is compared between LVCMOS 25, LVCMOS 33 IO standard, and LVCMOS 12, LVCMOS 15 IO standard, which is an increment of 0.619%. The TP consumption increment is 0.826% for LVCMOS 18 IO standard.

REFERENCES

1. Kumar, Keshav, Amanpreet Kaur, S. N. Panda, and Bishwajeet Pandey. "Effect of different nano meter technology-based FPGA on energy efficient UART design." In *2018 8th International Conference on Communication Systems and Network Technologies (CSNT)*, India, pp. 1–4. IEEE, 2018.
2. Kumar, Keshav, Amanpreet Kaur, Bishwajeet Pandey, and S. N. Panda. "Low power UART design using different nanometer technology based FPGA." In *2018 8th International Conference on Communication Systems and Network Technologies (CSNT)*, India, pp. 1–3. IEEE, 2018.
3. Kumar, Keshav, Bishwajeet Pandey, Amit Kant Pandit, Y. A. Baker El-Ebiary, Salameh A. Mjlae, and Samer Bamansoor. "Design of low power transceiver on Spartan-3 and Spartan-6 FPGA." *International Journal of Innovative Technology and Exploring Engineering* 8, no. 12S2 (2019): 27–30.
4. Basics of UART explained - Communication protocol and its applications. ElProCus. https://www.elprocus.com/basics-of-uart-communication-block-diagram-applications/. Accessed on 30 June 2023.
5. Sandhu, Amanpreet, Vidhoytma Gandhi, Simranpreet Kaur, Surbhi Huria, Divjot Singh, and Wamika Goyal. "Thermally aware LVCMOS based low power universal asynchronous receiver transmitter design on FPGA." *Indian Journal of Science and Technology* 8, no. 20 (2015): 1–4.

6. Kumar, Abhishek, Bishwajeet Pandey, D. M. Akbar Hussain, Mohammad Atiqur Rahman, Vishal Jain, and Ayoub Bahanasse. "Low voltage complementary metal oxide semiconductor based energy efficient UART design on Spartan-6 FPGA." In *2019 11th International Conference on Computational Intelligence and Communication Networks (CICN)*, India, pp. 84–87. IEEE, 2019.

7. Pandey, Bishwajeet, and Ravikant Kumar. "Low voltage DCI based low power VLSI circuit implementation on FPGA." In *2013 IEEE Conference on Information & Communication Technologies,* India, pp. 128–131. IEEE, 2013.

8. Kaur, Harkinder, Harsh Sohal, and Jaiteg Singh. "Design and performance analysis of UART using Altera Quartus-II and Xilinx ISE 14.2." In *6th International Conference on Communication and Network Technologies*, India. 2016.

9. Kaur, Ravinder, Jagdish Kumar, Sumita Nagah, Bishwajeet Pandey, and Kavita Goswami. "IO Standard based low power memory design and implementation on FPGA." In *2015 2nd International Conference on Computing for Sustainable Global Development (INDIACom)*, India, pp. 1501–1505. IEEE, 2015.

10. Sharma, Rashmi, Bishwajeet Pandey, Vikas Jha, Siddharth Saurabh, and Sweety Dabas. "Input–output standard-based energy efficient UART design on 90nm FPGA." In Sunil Kumar Muttoo (Ed.), *System and architecture*, pp. 139–150. Springer, Singapore, 2018.

11. Kumar, Keshav, Amanpreet Kaur, and K. R. Ramkumar. "Effective data transmission with UART on Kintex-7 FPGA." In *2020 12th International Conference on Computational Intelligence and Communication Networks (CICN)*, India, pp. 492–497. IEEE, 2020.

12. Aggarwal, Arushi, Bishwajeet Pandey, Sweety Dabbas, Achal Agarwal, and Siddharth Saurabh. "LVCMOS-based low-power thermal-aware energy-proficient vedic multiplier design on different FPGAs." In Sunil Kumar Muttoo (Ed.), *System and architecture*, pp. 115–122. Springer, Singapore, 2018.

13. Kumar, Abhishek, Bishwajeet Pandey, D. M. Akbar Hussain, Mohammad Atiqur Rahman, Vishal Jain, and Ayoub Bahanasse. "Frequency scaling and high speed transceiver logic based low power UART design on 45nm FPGA." In *2019 11th International Conference on Computational Intelligence and Communication Networks (CICN)*, Honolulu, HI, pp. 88–92. IEEE, 2019.

14. Gupta, Isha, Swati Singh Garima, Harpreet Kaur, Deepshikha Bhatt, and Aamir Vohra. "28nm FPGA based power optimized UART design using HSTL I/O standards." *Indian Journal of Science and Technology* 8, no. 17 (2015): 1–6.

15. Kumar, Keshav, Bishwajeet Pandey, and D. A. Hussain. "Power efficient UART design using capacitive load on different nanometer technology FPGA." *Gyancity Journal of Engineering and Technology* 5, no. 2 (2019): 1–13.

16. Kumar, Vivek, Aksh Rastogi, and V. K. Tomar. "Implementation of UART design for RF modules using different FPGA technologies." *IOP Conference Series: Materials Science and Engineering* 1116, no. 1 (2021): 012131.

17. Kamath, Akshatha, Tanya Mendez, S. Ramya, and Subramanya G. Nayak. "Design and implementation of power-efficient FSM based UART." *Journal of Physics: Conference Series* 2161, no. 1 (2022): 012052.

Chapter 5

SSTL-based UART of GCC

LIST OF ABBREVIATIONS

BUFG	Global Buffers
Clk	Clocks
DDR	Double Data Rate
DRAM	Dynamic Random-Access Memory
DP	Dynamic Power
FF	Flip-Flops
FPGA	Field Programmable Gate Array
GC	Green Computing
GCC	Green Communication Computing
G. Comm.	Green Communication
GPS	Global Positioning System
IC	Integrated Circuit
IO	Input Output
JT	Junction Temperature
LP	Leakage Power
LUT	Look Up Tables
RTL	Register Transfer Logic
SSTL	Stub Series Terminated Logic
S/G	Signal
SP	Static Power
TM	Thermal Margin
TP	Total Power
TPC	Total Power Consumption
UART	Universal Asynchronous Receiver Transmitter
θJA	Effective Thermal Resistance to Air

DOI: 10.1201/9781003302872-5

5.1 INTRODUCTION

A balance is always required in each and every process of environment. It is applicable for energy and power production and consumption as well. The deficiency in energy and power is the situation where the production of energy and power cannot cope with the consumption [1]. Therefore, this imbalance leads to energy crisis, which can be seen in the entire globe at this moment. An energy crisis is any significant bottleneck in the supply of energy resources to an economy. The causes of energy crisis are industrialization, shortage, less production, overpopulation, and urbanization. Apart from these causes, there is one major cause which is overconsumption [2]. There are many sectors in society that consume much energy than required. The energy crisis across the globe plays a key role in DE functioning of proper data transmission [3]. When there is no power, the data communication process gets stuck and there is a delay in data transmission. To achieve a proper communication, we should move toward a system that consumes less power. In order to achieve this, in this chapter, we are designing a low-power UART on Kintex-7 device with the help of stub series terminated logic (SSTL) IO standards. In Chapter 4, we have discussed about the use of IO standards, and about the working functionality of UART. With the help of IO standards, we can optimize the power consumption. Lower the power consumption, the lesser the energy production resources required. One can say that both goes hand in hand.

5.2 INTRODUCTION TO SSTL IO STANDARD

SSTL is the acronym for stub series terminated logic. These are a group of electrical interfaces used for matching the impedance of transmission line. These IO standards are commonly used with dynamic random-access memory (DRAM)-based double data rate (DDR) memory ICs. SSTL IO standards are used in many other applications too, especially in high-speed devices. SSTL IO standards are unidirectional and bidirectional [4,5]. There are four different levels of voltages in SSTL IO, and hence it can be classified into four categories as follows:

- SSTL 12: Operating input voltage of 1.2 V
- SSTL 135: Operating input voltage of 1.35 V
- SSTL 15: Operating input voltage of 1.5 V
- SSTL 18: Operating input voltage of 1.8 V

There are four distinguished classes of SSTL IO standards, and every class has its own unique features, which are represented in Table 5.1.

5.3 INTEGRATION OF UART WITH FPGA

This section discusses about the implementation of UART on FPGA. Here the UART is implemented on Kintex-7 FPGA device. The transmission line impedance matching is done with the help of SSTL IO standards. The register transfer logic (RTL) design is described in Figure 5.1.

In Figure 5.1, there are two UARTs: one is for transmitting the signal, and the other is for receiving the signal. At the input of transmitting end, there are five wires (clk, 16-bit perscale, rst, s_axis_tdata 8 bit, and s_axix_tvalid), while at output side there are three wires (busy, txd, and s_axix_tready). Similarly, at the input of receiving side, there are six wires (clk, 16-bit perscale, rst, m_axis_tready, rst, and rxd) while at output side, there are five wires (busy, frame_error, m_axis_tdata 8 bit, m_axis_tvalid, and overrun_error). The gate-level schematic circuit is illustrated in Figure 5.2, and the synthesized FPGA interfacing of UART is shown in Figure 5.3 [6–8].

Table 5.1 Classes of SSTL IO standards and features

Classes	Features
Class I	Unterminated, or symmetrically parallel terminated
Class II	Series terminated
Class III	Asymmetrically parallel terminated
Class IV	Asymmetrically parallel double terminated

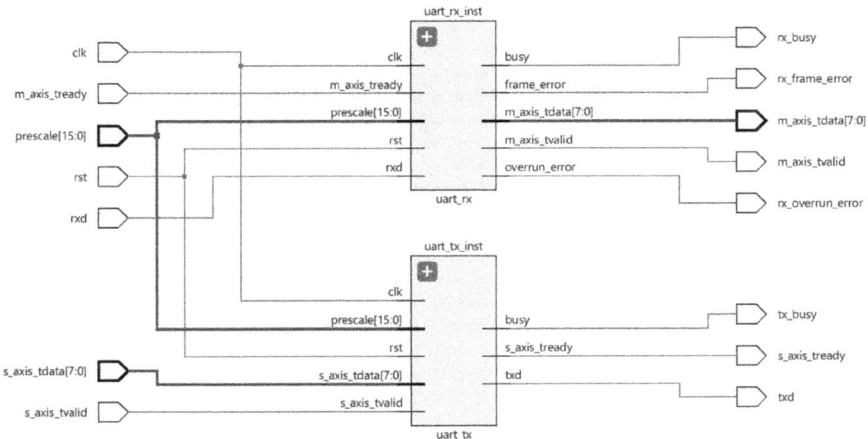

Figure 5.1 RTL design of UART.

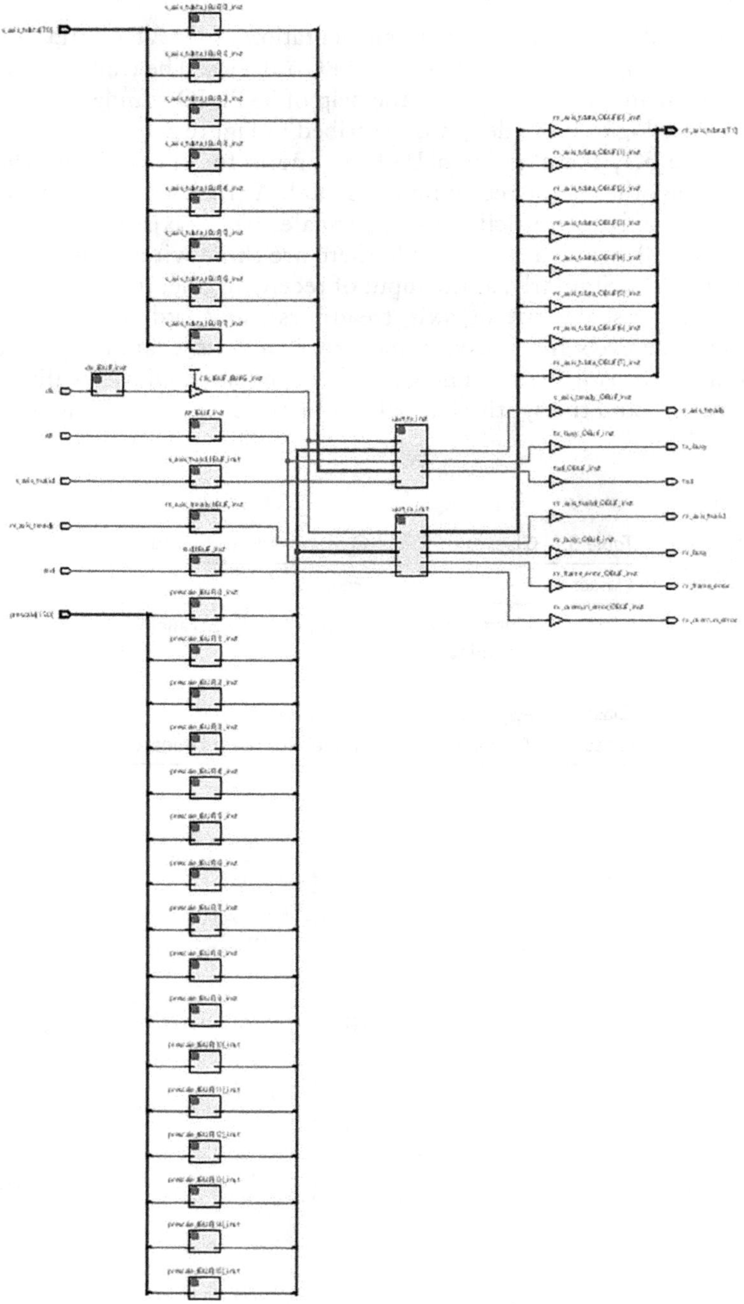

Figure 5.2 Gate-level schematic circuit.

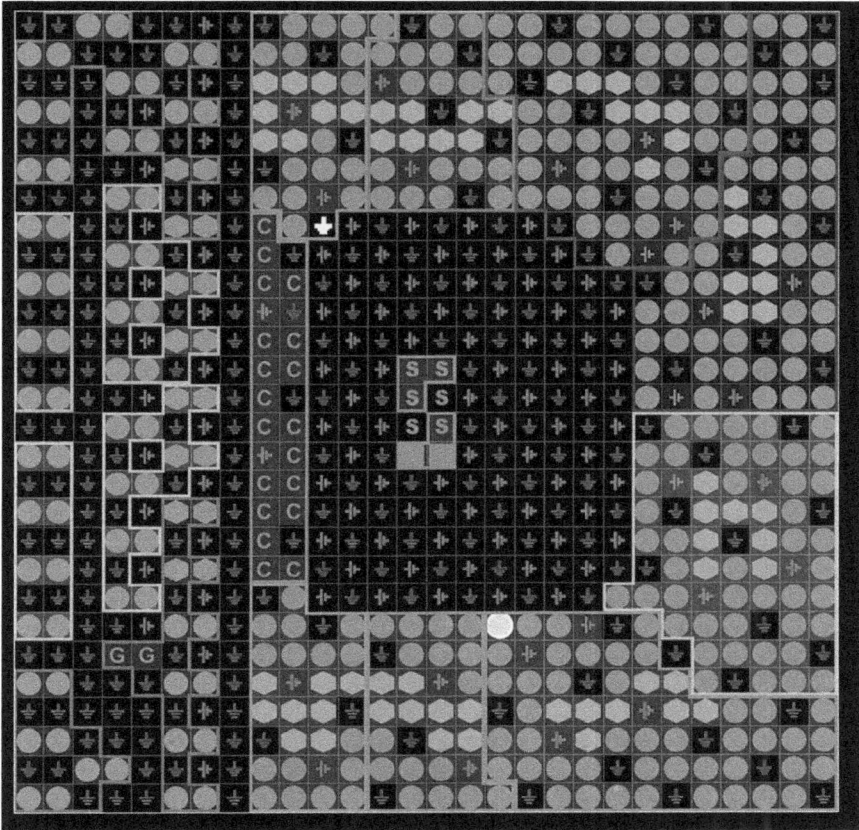

Figure 5.3 Synthesized FPGA interfacing of UART.

5.4 RESOURCE UTILIZATION

There are some FPGA resources used in the implementation of UART on Kintex-7 device. There resources are look up tables (LUTs), flip-flops (FF), input output (IO), and global buffers (BUFG). In the implementation of UART, Kintex-7 consumes 79 FFs, 110 LUTs, 44 IOs, and 1 BUFG. The resource utilization of Kintex-7 is described in Table 5.2, and the percentage of resource consumed is shown in Figure 5.4 [9–11].

Table 5.2 Resource utilization of UART on Kintex-7

Resources	Utilization	Available	Utilization %
FF	79	43,3920	0.02
IO	44	304	14.47
LUT	110	21,6960	0.05
BUFG	1	256	0.39

Utilization %

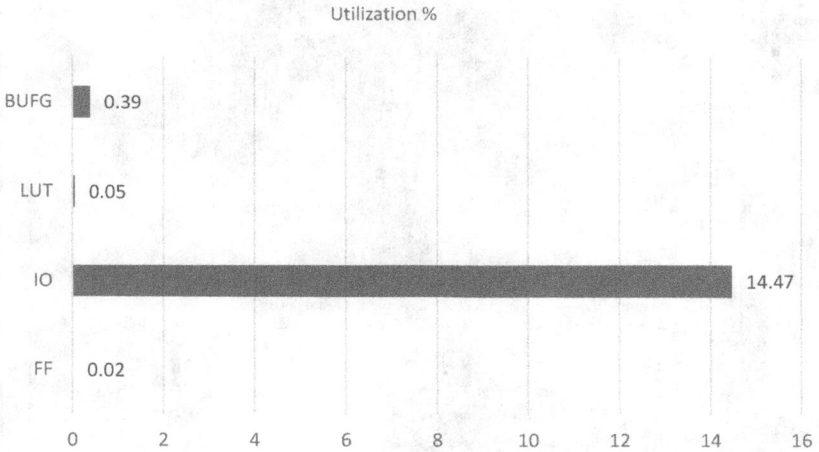

Figure 5.4 Percentage of resource consumed.

5.5 THERMAL PROPERTIES

This section describes the thermal characteristics of the Kintex-7 device when the impedance matching is done with the help of SSTL IO standards for UART implementation on FPGA. These thermal properties are junction temperature (JT), thermal margin (TM), and effective thermal resistance (ϑJA) [12–14].

a. JT: It is the highest operating temperature of the inbuilt gates in a FPGA device.
b. TM: it is the property of the devices which allows the device to consume low power.
c. ϑJA: It is defined as the amount of heat generated or a rise in temperature when 1 W of power is dissipated in the IC.

The thermal properties associated with SSTL IO standards for Kintex-7 device are described in Table 5.3.

Table 5.3 Thermal properties associated with SSTL IO standards for Kintex-7

SSTL IO	JT (°C)	TM (°C)	θJA (°C/W)
SSTL 12	25.3	74.7	1.9
SSTL 135	25.3	74.7	1.9
SSTL 15	25.3	74.7	1.9
SSTL 18_I	25.2	74.8	1.9
SSTL 18_II	25.2	74.8	1.9

From Table 5.3, it is observed that the JT, TM, and θJA are identical for SSTL 12, SSTL 135, and SSTL 15 IO standards. The JT and TM changes for SSTL 18_I and SSTL 18_II in comparison with SSTL 12, SSTL 135, and SSTL 15 IO standards. The θJA is the same for all the SSTL IOs.

5.6 POWER ANALYSIS

Power consumption is one of the most pressing concerns confronting the whole electronics industry. We're all working together to create a low-power system. Lowering the power usage improves the device's life system. In this section, we calculate the power consumption of a UART implementation on a Kintex-7 device by matching its impedance to the SSTL IO standard. The device's overall power consumption is the sum of both dynamic and static power as mathematically expressed in Equation 5.1 [15,16].

$$TP = SP + DP \qquad (5.1)$$

where

TP = Total power
SP = Static Power
DP = Dynamic Power

The device SP comprises the sum up of Clocks (clk), IO, Logic, and Signa (S/G) power, and the DP is the leakage power (LP) of the device.

5.6.1 Power analysis for SSTL 12

For the SSTL 12 IO, the SP of the device is 0.112 W which is 81% of the TP, and the DP is 0.026 W which is the 19% of the TP. The TP is the sum pf DP and SP, which is 0.138 W for SSTL 12. The TP consumption for SSTL 12 IO is represented in Figure 5.5.

On-Chip Power

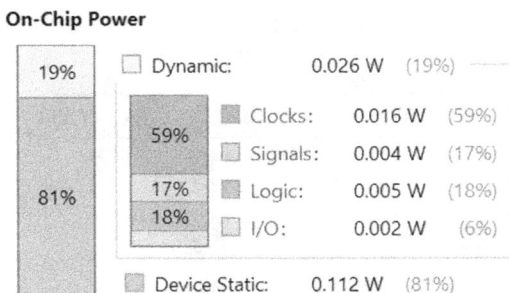

Figure 5.5 TP consumption for SSTL 12 IO.

TP (W) for SSTL 12

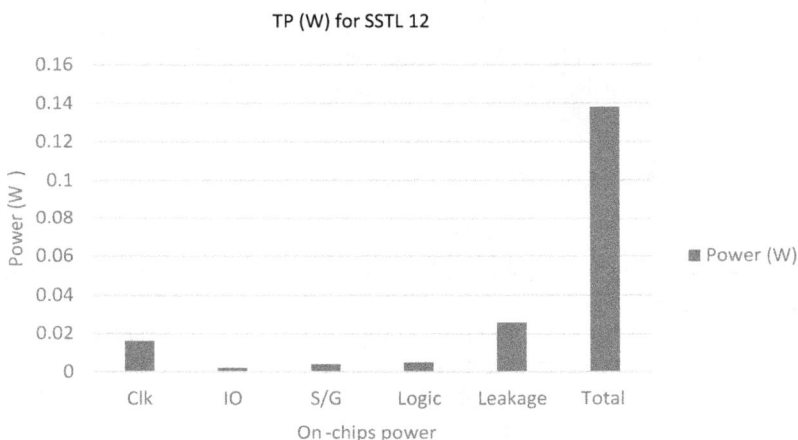

Figure 5.6 Graphical representation of on-chips power for SSTL 12 IO.

We know that the SP is the summation of clk, IO, S/G, and logic power. The clk, IO, S/G, and logic power are 0.016, 0.002, 0.004, and 0.005 W, respectively, for SSTL 12 IO. The leakage power is the DP which is 0.026 W. The graphical representation of on-chips power of the device is shown in Figure 5.6.

5.6.2 Power analysis for SSTL 135 and SSTL 15 IO

When the SSTL 135 and SSTL 15 IO standards are used for matching the impedance of the device, the TP consumption of the device observed is 0.139 W. The device SP is 0.112 W which is 81% of TP, and DP is 0.027 W which is 19% of TP consumption. The power consumption for SSTL 135 and SSTL 15 IO is illustrated in Figure 5.7. The graphical representation of TP consumption is described in Figure 5.8.

On-Chip Power

19%	Dynamic: 0.027 W (19%)
	Clocks: 0.016 W (58%)
58%	
	Signals: 0.004 W (16%)
81%	Logic: 0.005 W (18%)
16%	
18%	I/O: 0.002 W (8%)
	Device Static: 0.112 W (81%)

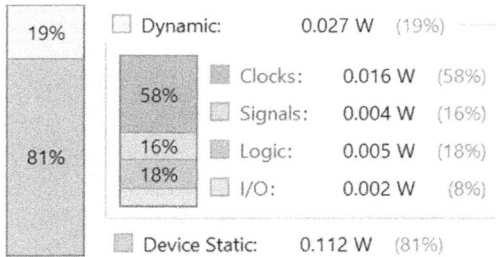

Figure 5.7 TP consumption for SSTL I35 and SSTL I5 IO.

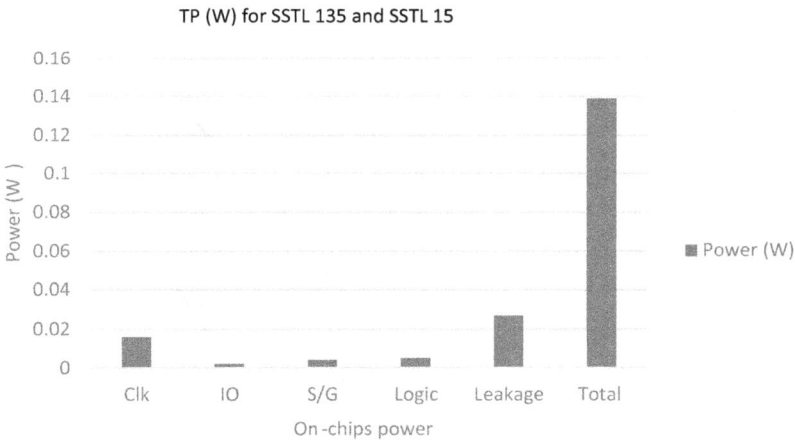

TP (W) for SSTL 135 and SSTL 15

Figure 5.8 Graphical representation of on-chips power for SSTL I35 and SSTL I5 IO.

From Figure 5.8, we know that the SP is the summation of clk, IO, S/G, and logic power. The clk, IO, S/G, and logic power are 0.016, 0.002, 0.004, and 0.005 W, respectively, for SSTL 12 IO. The leakage power is the DP which is 0.027 W.

5.6.3 Power analysis for SSTL 18_I and SSTL 18_II

For the SSTL 18_I and SSTL 18_II IO, the SP of the device is 0.112 W which is 95% of the TP, and the DP is 0.005 W which is the 5% of the TP. The TP is the sum of DP and SP, which is 0.117 W for SSTL 18_I and SSTL 18_II. The TP consumption for SSTL 12 IO is represented in Figure 5.9.

On-Chip Power

☐ Dynamic:	0.005 W	(5%)	
▨ Clocks:	0.003 W	(58%)	
☐ Signals:	0.001 W	(16%)	
▨ Logic:	0.001 W	(18%)	
☐ I/O:	<0.001 W	(8%)	
▨ Device Static:	0.112 W	(95%)	

95%   58%   16%   18%

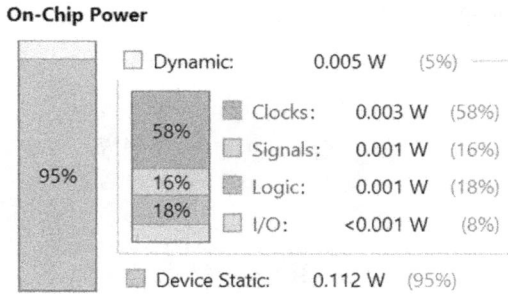

Figure 5.9 TP consumption for SSTL 18_I and SSTL 18_II IO.

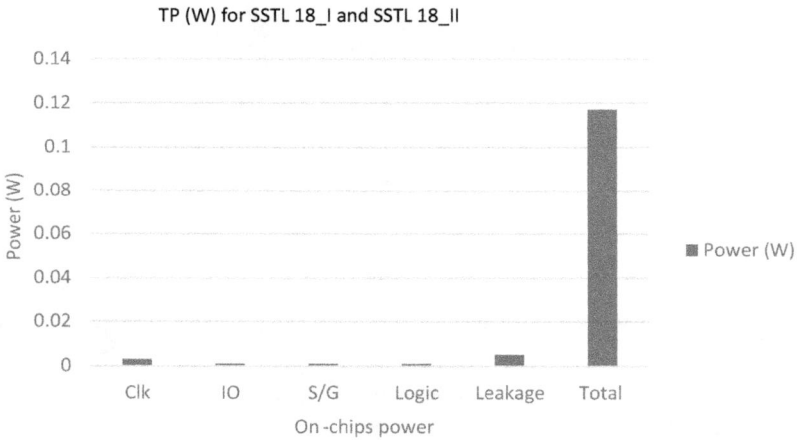

TP (W) for SSTL 18_I and SSTL 18_II

■ Power (W)

On-chips power

Clk   IO   S/G   Logic   Leakage   Total

Figure 5.10 Graphical representation of on-chips power for SSTL 18_I and SSTL 18_II IO.

We know that the SP is the summation of clk, IO, S/G, and logic power. The clk, IO, S/G, and logic power are 0.003, <0.001, 0.001, and 0.001S W, respectively, for SSTL 18_I and SSTL 18_II IO. The leakage power is the DP which is 0.005 W. The graphical representation of on-chips power of the device is shown in Figure 5.10.

5.7 OBSERVATION AND COMPARATIVE ANALYSIS

From the power analysis section, it is observed that the SP of the device remains the same for all the SSTL IOs. The change is observed in the DP of the device. The switch in the DP causes the switch in the TP consumption. For SSTL 12 IO, the TP is 0.138 W; for the SSTL 135 and SSTL 15, the TP remains identical, i.e., 0.139 W; and for the SSTL 18_I and SSTL

Table 5.4 TP consumption for used IO standards

IO standards	TP (W)
SSTL 12	0.138
SSTL 135	0.139
SSTL 15	0.139
SSTL 18_I	0.117
SSTL 18_II	0.117

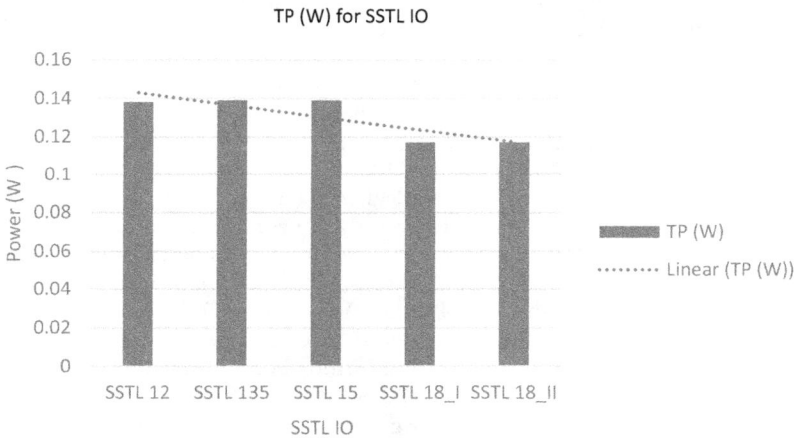

Figure 5.11 TP consumption for used IO standards.

18_II, the TP is 0.117 W. The TP consumption for the used IO standards is shown in Table 5.4 and illustrated in Figure 5.11. From the power analysis, it is clearly observed that for getting low power consumption of UART on Kintex-7 device, one should match the impedance of the device with SSTL 18_I and SSTL 18_II IO standards.

5.8 CONCLUSION

It has been noted that the world has been suffering a major problem of energy and power deficit in recent years. The earth's natural resources may not be around for long. It must finish at some point. The concept of GC comes to mind in the context of such a massive problem. This chapter represents a step forward in the GCC sector. In this chapter, a low-power UART model is created using the Kintex-7 device. In this chapter, the authors utilized the impedance matching approach with the SSTL IO standards to optimize the power usage of UART. FPGA IO standards are utilized to match the circuit's input and output impedance. It is observed that the TP consumption is optimized with the use of SSTL 18_I and SSTL 18_II IO standards. The TP consumption is maximum for SSTL 135 and SSTL 15 IO standards. The change in percentage increment from the optimum TP consumption to maximum TP consumption is 18.80%.

REFERENCES

1. Gandotra, Pimmy, Rakesh Kumar Jha, and Sanjeev Jain. "Green communication in next generation cellular networks: A survey." *IEEE Access* 5 (2017): 11727–11758.
2. Mahapatra, Rajarshi, Yogesh Nijsure, Georges Kaddoum, Naveed Ul Hassan, and Chau Yuen. "Energy efficiency tradeoff mechanism towards wireless green communication: A survey." *IEEE Communications Surveys & Tutorials* 18, no. 1 (2015): 686–705.
3. Vereecken, Willem, Ward Van Heddeghem, Didier Colle, Mario Pickavet, and Piet Demeester. "Overall ICT footprint and green communication technologies." In *2010 4th International Symposium on Communications, Control and Signal Processing (ISCCSP)*, Limassol, Cyprus, pp. 1–6. IEEE, 2010.
4. Sharma, Rashmi, Bishwajeet Pandey, Vikas Jha, Siddharth Saurabh, and Sweety Dabas. "Input–output standard-based energy efficient UART design on 90nm FPGA." In Sunil Kumar Muttoo (Ed.), *System and architecture*, pp. 139–150. Springer, Singapore, 2018.
5. 7 series FPGAs configuration user guide (UG470). https://www.eng.auburn.edu/~nelson/courses/elec4200/FPGA/ug470_7Series_Config.pdf. Accessed on 30 June 2023.
6. Kumar, Keshav, Amanpreet Kaur, S. N. Panda, and Bishwajeet Pandey. "Effect of different nano meter technology-based FPGA on energy efficient UART design." In *2018 8th International Conference on Communication Systems and Network Technologies (CSNT)*, India, pp. 1–4. IEEE, 2018.
7. Kumar, Keshav, Amanpreet Kaur, Bishwajeet Pandey, and S. N. Panda. "Low power UART design using different nanometer technology based FPGA." In *2018 8th International Conference on Communication Systems and Network Technologies (CSNT)*, India, pp. 1–3. IEEE, 2018.
8. Kumar, Keshav, Bishwajeet Pandey, Amit Kant Pandit, Y. A. Baker El-Ebiary, Salameh A. Mjlae, and Samer Bamansoor. "Design of low power transceiver on Spartan-3 and Spartan-6 FPGA." *International Journal of Innovative Technology and Exploring Engineering* 8, no. 12S2 (2019): 27–30.
9. Sandhu, Amanpreet, Vidhoytma Gandhi, Simranpreet Kaur, Surbhi Huria, Divjot Singh, and Wamika Goyal. "Thermally aware LVCMOS based low power universal asynchronous receiver transmitter design on FPGA." *Indian Journal of Science and Technology* 8, no. 20 (2015): 1–4.
10. Kumar, Abhishek, Bishwajeet Pandey, D. M. Akbar Hussain, Mohammad Atiqur Rahman, Vishal Jain, and Ayoub Bahanasse. "Low voltage complementary metal oxide semiconductor based energy efficient UART design on Spartan-6 FPGA." In *2019 11th International Conference on Computational Intelligence and Communication Networks (CICN)*, India, pp. 84–87. IEEE, 2019.

11. Pandey, Bishwajeet, and Ravikant Kumar. "Low voltage DCI based low power VLSI circuit implementation on FPGA." In *2013 IEEE Conference on Information & Communication Technologies,* India, pp. 128–131. IEEE, 2013.

12. Kaur, Harkinder, Harsh Sohal, and Jaiteg Singh. "Design and performance analysis of uart using Altera Quartus-II and Xilinx ISE 14.2." In *6th International Conference on Communication and Network Technologies,* India. 2016.

13. Kaur, Ravinder, Jagdish Kumar, Sumita Nagah, Bishwajeet Pandey, and Kavita Goswami. "IO Standard based low power memory design and implementation on FPGA." In *2015 2nd International Conference on Computing for Sustainable Global Development (INDIACom),* India, pp. 1501–1505. IEEE, 2015.

14. Sharma, Rashmi, Bishwajeet Pandey, Vikas Jha, Siddharth Saurabh, and Sweety Dabas. "Input–output standard-based energy efficient UART design on 90nm FPGA." In Sunil Kumar Muttoo (Ed.), *System and architecture,* pp. 139–150. Springer, Singapore, 2018.

15. Kumar, Keshav, Amanpreet Kaur, and K. R. Ramkumar. "Effective data transmission with UART on Kintex-7 FPGA." In *2020 12th International Conference on Computational Intelligence and Communication Networks (CICN),* India, pp. 492–497. IEEE, 2020.

16. Aggarwal, Arushi, Bishwajeet Pandey, Sweety Dabbas, Achal Agarwal, and Siddharth Saurabh. "LVCMOS-based low-power thermal-aware energy-proficient vedic multiplier design on different FPGAs." In Sunil Kumar Muttoo (Ed.), *System and architecture,* pp. 115–122. Springer, Singapore, 2018.

Chapter 6

HSTL-based UART for GCC

LIST OF ABBREVIATIONS

ASIC	Application-Specific Integrated Circuit
BUFG	Global Buffers
Clk	Clocks
CMOS	Complementary Metal-Oxide Semiconductor
DP	Dynamic Power
FF	Flip-Flops
FPGA	Field Programmable Gate Array
GC	Green Computing
GCC	Green Communication Computing
G. Comm.	Green Communication
HSTL	High-Speed Transceiver Logic
IC	Integrated Circuit
IO	Input Output
JT	Junction Temperature
LP	Leakage Power
LUT	Look Up Tables
RTL	Register Transfer Logic
S/G	Signal
SP	Static Power
TM	Thermal Margin
TP	Total Power
TPC	Total Power Consumption
UART	Universal Asynchronous Receiver Transmitter
ϑJA	Effective Thermal Resistance to Air

6.1 INTRODUCTION

The energy crisis refers to the fear that the world's demands on the finite natural resources required to power modern civilization are shrinking as the demand increases. Natural resources are in short supply. While they do occur naturally, replenishing the stockpiles can take hundreds of thousands of years. Governments and concerned citizens are collaborating to prioritize the use of renewable resources and reduce reckless use of natural resources through enhanced conservation [1]. The energy problem is a big and complicated subject. Most people aren't aware of its existence unless the price of petrol at the pump rises or there are long queues at the gas station. Despite several attempts, the energy crisis is still ongoing and worsening. The reason for this is because there is a lack of general awareness of the complicated reasons and answers to the energy issue, which will enable for a resolution attempt to take place [2,3].

There are various reasons that contribute to the global energy crisis, including:

a. Overconsumption
b. Overpopulation
c. Poor Infrastructure
d. Unexplored Renewable Energy Options
e. Delay in Commissioning of Power Plants
f. Wastage of Energy
g. Poor Distribution System
h. Major Accidents and Natural Calamities
i. Wars and Attacks

As a result, in order to accomplish proper communication, we need to shift to a system that uses less power. In order to do this, we will develop a low-power UART on a Kintex-7 device using high-speed transceiver logic (HSTL) IO standards in this chapter [4]. We have analyzed the use of IO standards and the operational capability of UART in Chapter 4. We can optimize power usage with the aid of IO standards. Reduced power use necessitates less energy producing resources. Both can be said to go hand in hand [5].

6.2 INTRODUCTION TO HSTL IO STANDARD

High-speed transceiver logic or HSTL is a technology-independent standard for signaling between integrated circuits. The nominal signaling range is from 0 to 1.5 V, though variations are allowed, and signals may be single-ended or differential. It is designed for operation beyond 180 MHz. HSTL can be implemented in both single-ended and differential forms and is intended to be a technology-independent standard, suitable for use with CMOS and Bipolar ICs [6,7].

The following classes are defined by standard EIA/JESD8-6 from EIA/JEDEC:

a. Class I (unterminated, or symmetrically parallel terminated)
b. Class II (series terminated)

c. Class III (asymmetrically parallel terminated)

d. Class IV (asymmetrically double parallel terminated)

6.3 INTEGRATION OF UART WITH FPGA

This section discusses about the implementation of UART on FPGA. Here the UART is implemented on Kintex-7 FPGA device. The transmission line impedance matching is done with the help of HSTL IO standards. The register transfer logic (RTL) design is described in Figure 6.1.

In Figure 6.1, there are two UARTs: one is for transmitting the signal, and the other is for receiving the signal. At the input of transmitting end, there are five wires (clk, 16-bit perscale, rst, s_axis_tdata 8 bit, and s_axix_tvalid), while at output side, there are three wires (busy, txd, and s_axix_tready). Similarly, at the input of receiving side, there are six wires (clk, 16-bit perscale, rst, m_axis_tready, rst, and rxd), while at output side, there are five wires (busy, frame_error, m_axis_tdata 8 bit, m_axis_tvalid, and overrun_error). The gate-level schematic circuit is illustrated in Figure 6.2, and the synthesized FPGA interfacing of UART is shown in Figure 6.3 [8–10].

6.4 RESOURCE UTILIZATION

There are some FPGA resources used in the implementation of UART on Kintex-7 device. The resources are look up tables (LUTs), flip-flops (FF),

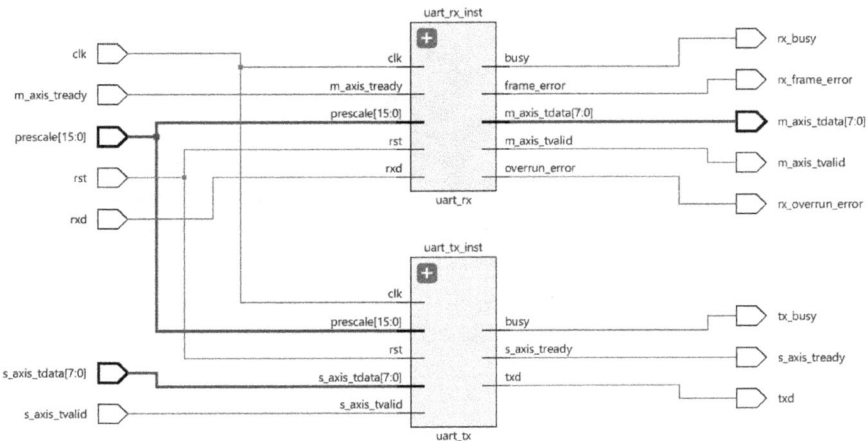

Figure 6.1 RTL design of UART.

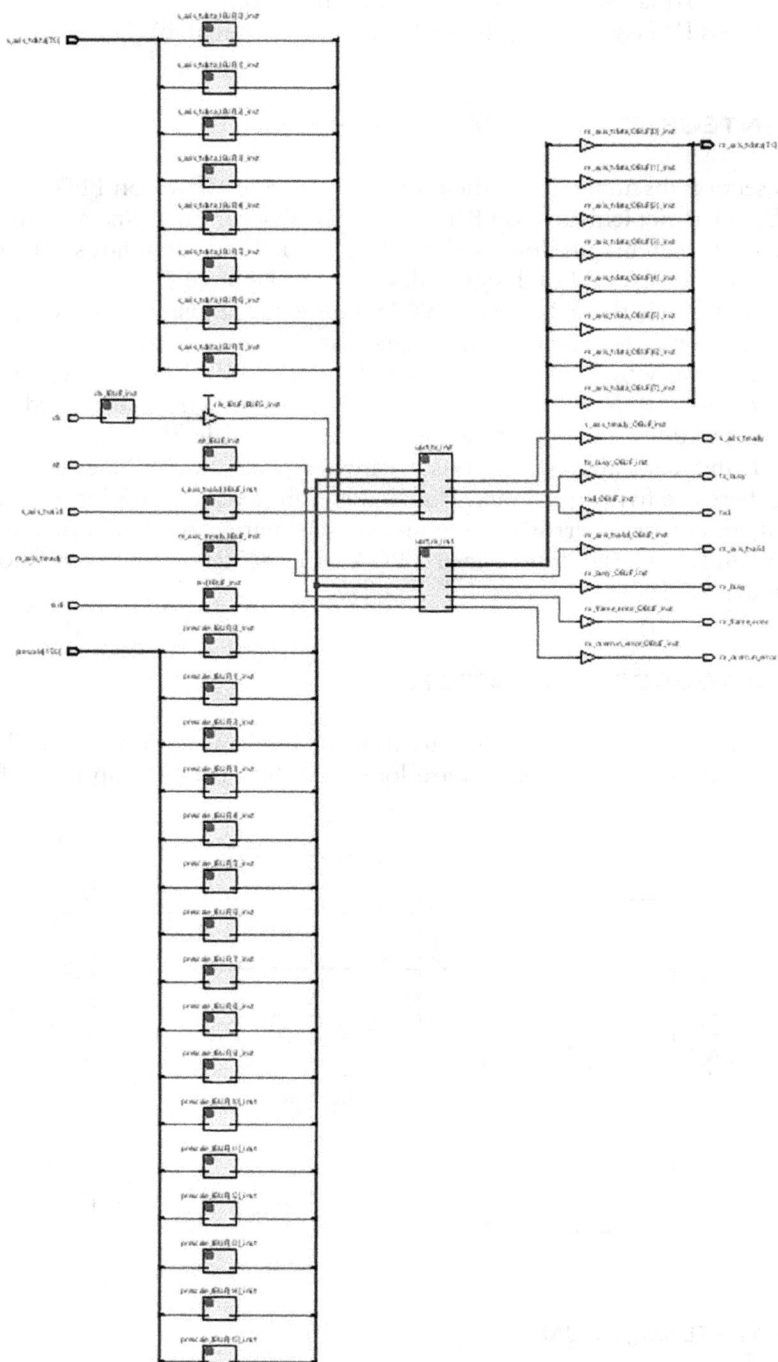

Figure 6.2 Gate-level schematic circuit.

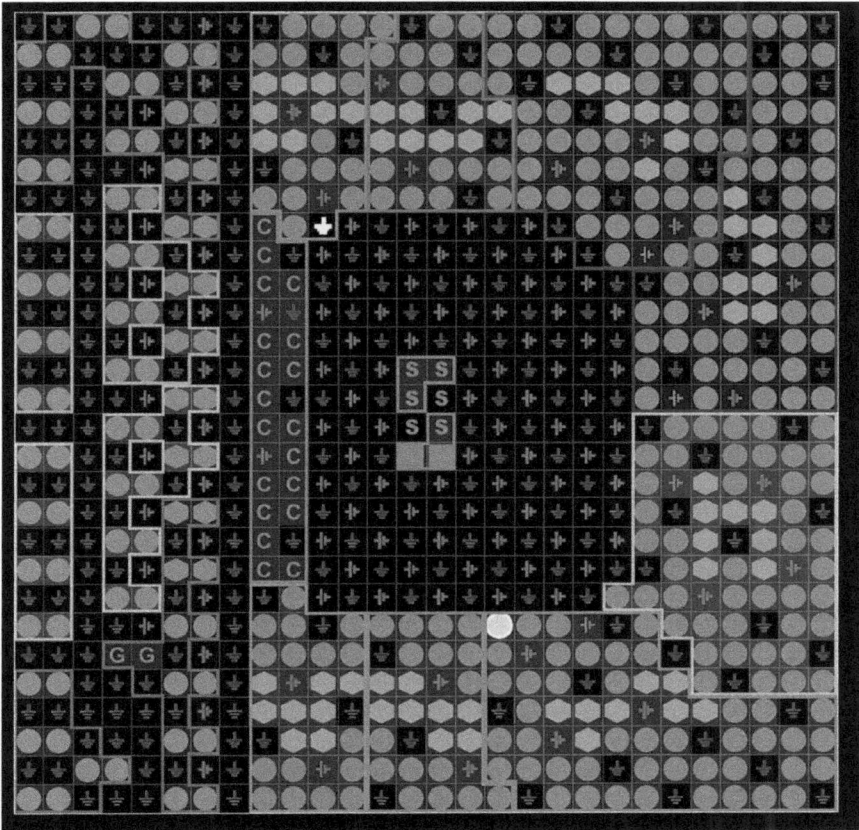

Figure 6.3 Synthesized FPGA interfacing of UART.

Table 6.1 Resource utilization of UART on Kintex-7

Resources	Utilization	Available	Utilization %
FF	79	43,3920	0.02
IO	44	304	14.47
LUT	110	21,6960	0.05
BUFG	1	256	0.39

input output (IO), and global buffers (BUFG). In the implementation of UART, Kintex-7 consumes 79 FFs, 110 LUTs, 44 IOs, and 1 BUFG. The resource utilization of Kintex-7 is described in Table 6.1, and the percentage of resource consumed is shown in Figure 6.4 [11–13].

Utilization %

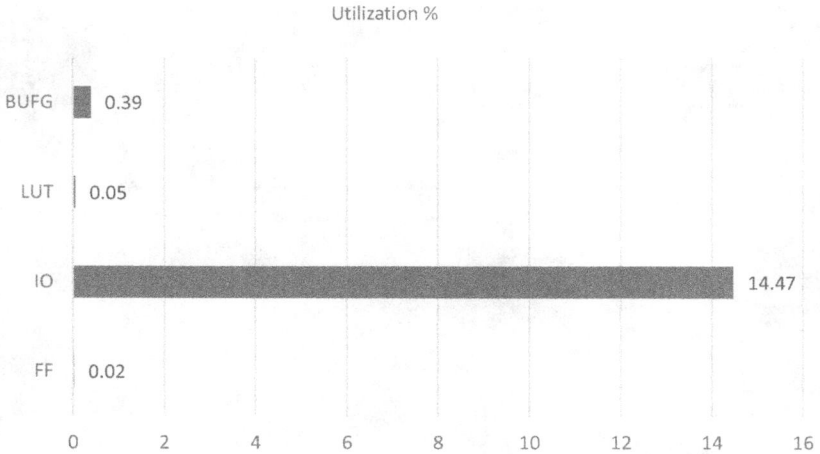

Figure 6.4 Percentage of resource consumed.

6.5 THERMAL PROPERTIES

This section describes the thermal characteristics of the Kintex-7 device when the impedance matching is done with the help of HSTL IO standards for UART implementation on FPGA. These thermal properties are junction temperature (JT), thermal margin (TM), and effective thermal resistance (ϑJA) [14–16].

a. JT: It is the highest operating temperature of the inbuilt gates in a FPGA device.
b. TM: It is the property of the devices which allows the device to consume low power.
c. ϑJA: It is defined as the amount of heat generated or a rise in temperature when 1 W of power is dissipated in the IC.

The thermal properties associated with HSTL IO standards for Kintex-7 device are described in Table 6.2.

From Table 6.2, it is observed that the thermal properties are identical for all the used HSTL IOs except HSTL_II_18 IO. For HSTL_II_18, the ϑJA is still the same, and there is only change in JT and TM.

6.6 POWER ANALYSIS

One of the most critical issues plaguing the whole electronics industry is power usage. We're all collaborating to develop a low-power system. Lowering power consumption enhances the device's life system. We calculate the power consumption of a UART implementation on a Kintex-7

Table 6.2 Thermal properties associated with HSTL IO standards for Kintex-7

HSTL IO	JT (°C)	TM (°C)	ϑJA (°C/W)
HSTL_I	28.0	72.0	1.9
HSTL_I_12	28.0	72.0	1.9
HSTL_I_18	28.0	72.0	1.9
HSTL_II	28.0	72.0	1.9
HSTL_II_18	28.1	71.9	1.9

device in this part by matching its impedance to the SSTL IO standard. The total power consumption of the gadget is the sum of dynamic and static power. Equation 6.1 expresses this mathematically [17–19].

$$TP = SP + DP \tag{6.1}$$

where

TP = Total Power
SP = Static Power
DP = Dynamic Power

The device SP is the total of the device's Clocks (clk), IO, Logic, and Signa (S/G) power, while the DP is the device's leakage power (LP).

6.6.1 Power analysis for HSTL_I

When the impedance matching is done with the help of HSTL_I IO, it is observed that the total power consumption is 1.588 W, which is the summation of DP (1.469 W) which is 93% of TP and SP is (0.119 W) which is 7% of TP. The TP consumption for HSTL_I IO is illustrated in Figure 6.5.

We know that the DP is the summation of IO, S/G, and logic power. The IO, S/G, and logic power are 0.528, 0.455, and 0.486 W, respectively, for HSTL_I IO. The leakage power is the SP which is 1.469 W. The graphical representation of on-chips power of the device is shown in Figure 6.6.

6.6.2 Power analysis for HSTL_II

When the HSTL_II IO standards are used for matching the impedance of the device, the TP consumption of the device observed is 1.61 W. The device SP is 0.119 W which is 7% of TP, and DP is 1.491 W which is 93% of TP consumption. The power consumption for HSTL_II IO is illustrated in Figure 6.7. The graphical representation of TP consumption is described in Figure 6.8.

On-Chip Power

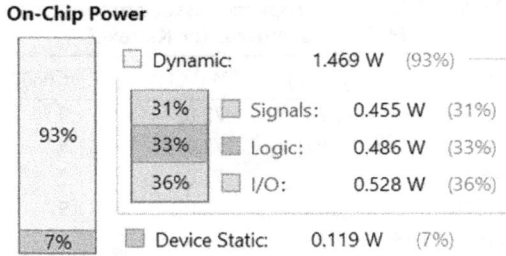

	Dynamic:	1.469 W	(93%)
31%	Signals:	0.455 W	(31%)
33%	Logic:	0.486 W	(33%)
36%	I/O:	0.528 W	(36%)
7%	Device Static:	0.119 W	(7%)

Figure 6.5 TP consumption for HSTL_I IO.

TP (W) for HSTL_I

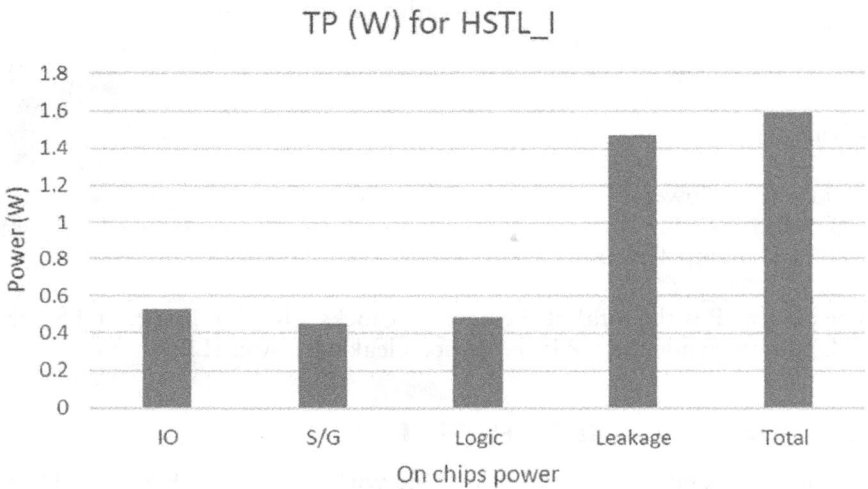

Figure 6.6 Graphical representation of on-chips power for HSTL_I IO.

On-Chip Power

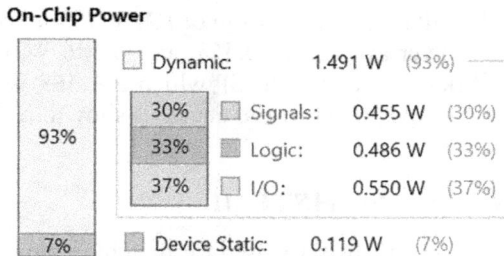

	Dynamic:	1.491 W	(93%)
30%	Signals:	0.455 W	(30%)
33%	Logic:	0.486 W	(33%)
37%	I/O:	0.550 W	(37%)
7%	Device Static:	0.119 W	(7%)

Figure 6.7 TP consumption for HSTL_II IO.

TP (W) for HSTL_II

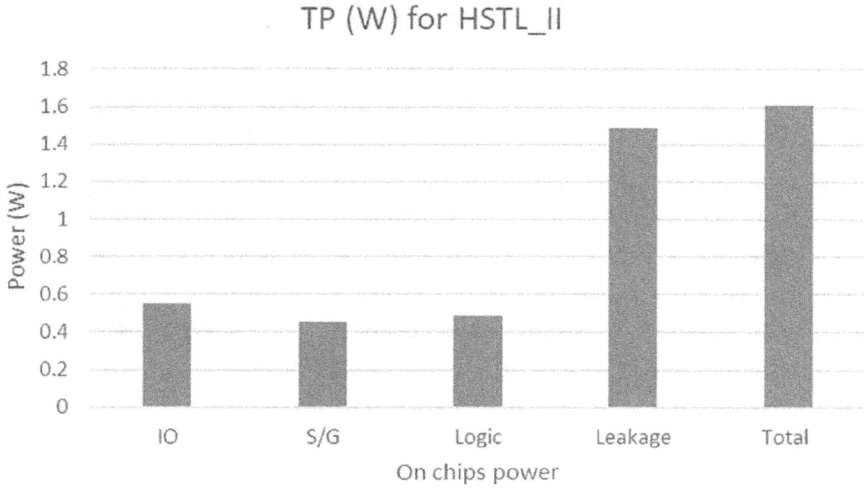

Figure 6.8 Graphical representation of on-chips power for HSTL_II IO.

On-Chip Power

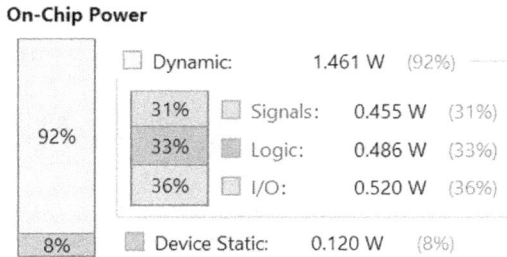

	Dynamic:	1.461 W	(92%)
	Signals:	0.455 W	(31%)
	Logic:	0.486 W	(33%)
	I/O:	0.520 W	(36%)
	Device Static:	0.120 W	(8%)

Figure 6.9 TP consumption for HSTL_I_I2 IO.

From Figure 6.8, we know that the DP is the summation of IO, S/G, and logic power. The IO, S/G, and logic power are 0.550, 0.455, and 0.486 W, respectively, for HSTL_II IO. The leakage power is the SP which is 1.491 W.

6.6.3 Power analysis for HSTL_I_I2

For the HSTL_I_12 IO, the SP of the device is 0.120 W which is 8% of the TP, and the DP is 1.461 W which is 92% of the TP. The TP is the sum pf DP and SP, which is 1.581 W. The TP consumption for HSTL_I_12 IO is represented in Figure 6.9.

We know that the DP is the summation of IO, S/G, and logic power. The IO, S/G, and logic power are 0.520, 0.455, and 0.486 W, respectively, for HSTL_I IO. The leakage power is the SP which is 1.461 W. The graphical representation of on-chips power of the device is shown in Figure 6.10.

TP (W) for HSTL_I_12

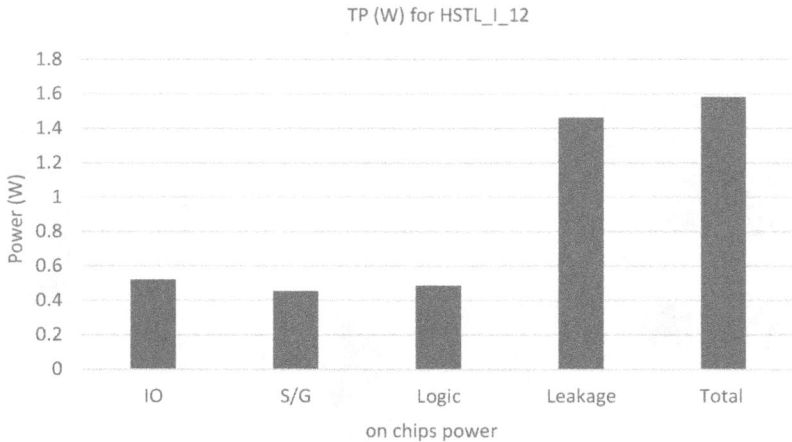

Figure 6.10 Graphical representation of on-chips power for HSTL_I_12 IO.

6.6.4 Power analysis for HSTL_I_18

When the impedance matching is done with the help of HSTL_I IO, it is observed that the total power consumption is 1.598 W, which is the summation of DP (1.479 W) which is 93% of TP and SP is (0.120 W) which is 7% of TP. The TP consumption for HSTL_I_18 IO is illustrated in Figure 6.11.

We know that the DP is the summation of IO, S/G, and logic power. The IO, S/G, and logic power are 0.538, 0.455, and 0.486 W, respectively, for HSTL_I IO. The leakage power is the SP which is 1.469 W. The graphical representation of on-chips power of the device is shown in Figure 6.12.

6.6.5 Power analysis for HSTL_II_18

When the HSTL_II_18IO standards are used for matching the impedance of the device, the TP consumption of the device observed is 1.629 W. The device SP is 0.119 W which is 7% of TP, and DP is 1.511 W which is 93% of TP consumption. The power consumption for HSTL_II_18 IO is illustrated in Figure 6.13. The graphical representation of TP consumption is described in Figure 6.14.

6.7 OBSERVATION AND COMPARATIVE ANALYSIS

This section gives a brief idea about the comparison of TPC of the device for the distinguished HSTL IO used for matching the impedance. From Section 6.6, it is observed that the device works under the category of low power consumption when the impedance is matched with HSTL_I_12 IO,

On-Chip Power

☐ Dynamic:	1.479 W (93%)
■ Signals:	0.455 W (31%)
■ Logic:	0.486 W (33%)
☐ I/O:	0.538 W (36%)
■ Device Static:	0.120 W (7%)

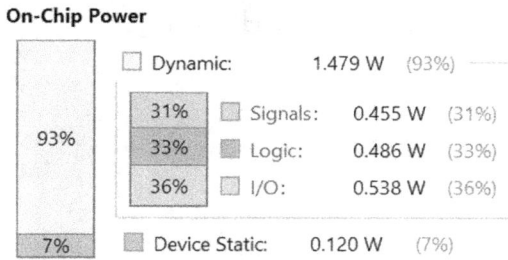

Figure 6.11 TP consumption for HSTL_I_18 IO.

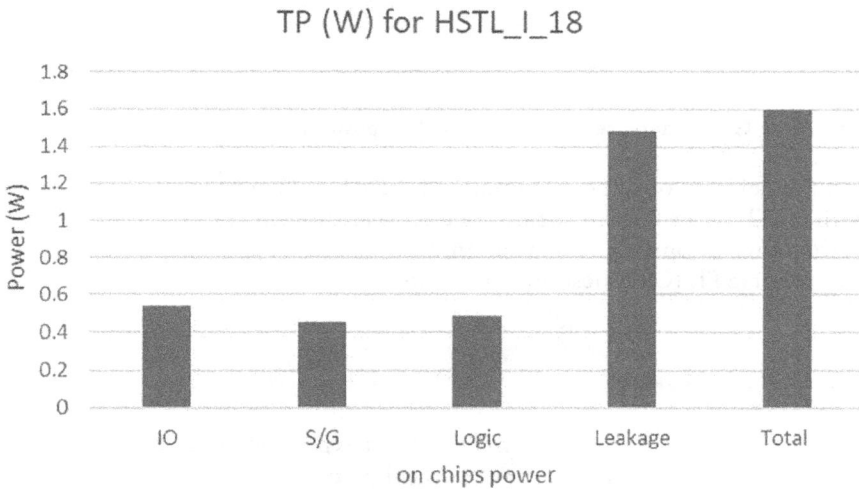

TP (W) for HSTL_I_18

Figure 6.12 Graphical representation of on-chips power for HSTL_I_18 IO.

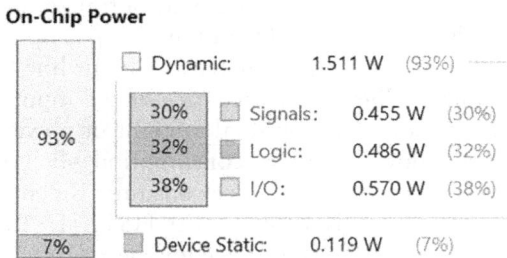

On-Chip Power

☐ Dynamic:	1.511 W (93%)
☐ Signals:	0.455 W (30%)
■ Logic:	0.486 W (32%)
☐ I/O:	0.570 W (38%)
■ Device Static:	0.119 W (7%)

Figure 6.13 TP consumption for HSTL_II_18 IO.

TP (W) for HSTL_II_18

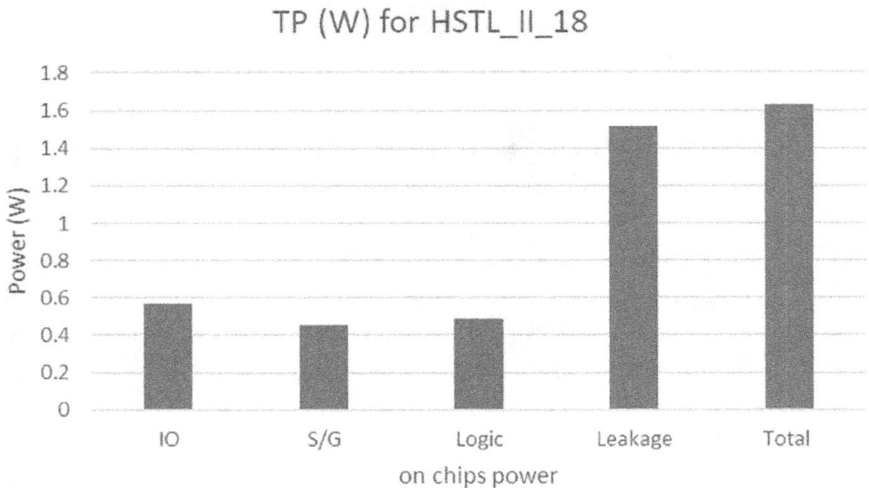

Figure 6.14 Graphical representation of on-chips power for HSTL_II_18 IO.

and the devices consume maximum power when the impedance is matched with HSTl_II_18 IO. For the rest of the other three HSTL IOs, the TPC lies in the range of optimal to maximum power. The TPC of the device for the different HSTL IOs is described in Figure 6.15.

6.8 CONCLUSION

It has been reported that the world has been experiencing severe energy and power shortages in recent years. Natural resources on the earth may not last long. It must vanish at some point. The concept of GC comes to mind in the context of such a massive problem. This chapter is a step forward in the GCC sector. In this chapter, a low-power UART model is created using the Kintex-7 device. The authors of this chapter utilized the impedance matching approach with the HSTL IO standards to optimize the power utilization of UART. In this chapter, the TPC of the device is optimized using the HSTL IO. IO standards are used to match the input line impedance with output line impedance. When the impedance of the input and the output is matched, the device consumes optimal amount of power. In the case of HSTL IO, the devices allow low power consumption when the impedance is matched with HSTL_I_12 IO. The device consumes the maximum amount of power when the impedance is matched with HSTL_II_18 IO. There is an increment of 3.036% in TPC of the device for the HSTL_I_12 (optimized) and HSTL_II_18 (maximum) IOs. The other three HSTL IOs consume power in between HSTL_I_12 and HSTL_II_18 IO. As far as the future

TPC (W)

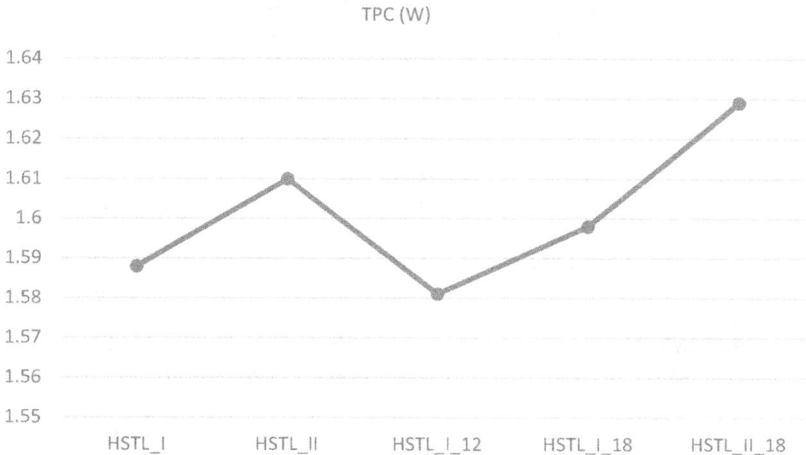

Figure 6.15 TPC of UART for HSTL IOs.

scope is concerned, the UART design can be implemented with the other SoC-based FPGAs, and there are also different power optimization techniques which can be used such as capacitance scaling, frequency variation of the device, and clock gating. Also, the FPGA design can be converted into the ASIC design for better results.

REFERENCES

1. Gandotra, Pimmy, Rakesh Kumar Jha, and Sanjeev Jain. "Green communication in next generation cellular networks: A survey." *IEEE Access* 5 (2017): 11727–11758.
2. Mahapatra, Rajarshi, Yogesh Nijsure, Georges Kaddoum, Naveed Ul Hassan, and Chau Yuen. "Energy efficiency tradeoff mechanism towards wireless green communication: A survey." *IEEE Communications Surveys & Tutorials* 18, no. 1 (2015): 686–705.
3. Vereecken, Willem, Ward Van Heddeghem, Didier Colle, Mario Pickavet, and Piet Demeester. "Overall ICT footprint and green communication technologies." In *2010 4th International Symposium on Communications, Control and Signal Processing (ISCCSP)*, Limassol, Cyprus, pp. 1–6. IEEE, 2010.
4. Gupta, Isha, Swati Singh Garima, Harpreet Kaur, Deepshikha Bhatt, and Aamir Vohra. "28nm FPGA based power optimized UART design using HSTL I/O standards." *Indian Journal of Science and Technology* 8, no. 17 (2015): 1–6.
5. Sharma, Rashmi, Bishwajeet Pandey, Vikas Jha, Siddharth Saurabh, and Sweety Dabas. "Input–output standard-based energy efficient UART design on 90nm FPGA." In Sunil Kumar Muttoo (Ed.), *System and architecture*, pp. 139–150. Springer, Singapore, 2018.

6. Gupta, Isha, Swati Singh Garima, Harpreet Kaur, Deepshikha Bhatt, and Aamir Vohra. "28nm FPGA based power optimized UART design using HSTL I/O standards." *Indian Journal of Science and Technology* 8, no. 17 (2015): 1–6.

7. Xilinx manual. 7 series FPGAs configuration user guide (UG470). https://www.eng.auburn.edu/~nelson/courses/elec4200/FPGA/ug470_7Series_Config.pdf. Accessed on 30 June 2023.

8. Kumar, Keshav, Amanpreet Kaur, S. N. Panda, and Bishwajeet Pandey. "Effect of different nano meter technology-based FPGA on energy efficient UART design." In *2018 8th International Conference on Communication Systems and Network Technologies (CSNT)*, India, pp. 1–4. IEEE, 2018.

9. Kumar, Keshav, Amanpreet Kaur, Bishwajeet Pandey, and S. N. Panda. "Low power UART design using different nanometer technology based FPGA." In *2018 8th International Conference on Communication Systems and Network Technologies (CSNT)*, India, pp. 1–3. IEEE, 2018.

10. Kumar, Keshav, Bishwajeet Pandey, Amit Kant Pandit, Y. A. Baker El-Ebiary, Salameh A. Mjlae, and Samer Bamansoor. "Design of low power transceiver on Spartan-3 and Spartan-6 FPGA." *International Journal of Innovative Technology and Exploring Engineering* 8, no. 12S2 (2019): 27–30.

11. Sandhu, Amanpreet, Vidhoytma Gandhi, Simranpreet Kaur, Surbhi Huria, Divjot Singh, and Wamika Goyal. "Thermally aware LVCMOS based low power universal asynchronous receiver transmitter design on FPGA." *Indian Journal of Science and Technology* 8, no. 20 (2015): 1–4.

12. Kumar, Abhishek, Bishwajeet Pandey, D. M. Akbar Hussain, Mohammad Atiqur Rahman, Vishal Jain, and Ayoub Bahanasse. "Low voltage complementary metal oxide semiconductor based energy efficient UART design on Spartan-6 FPGA." In *2019 11th International Conference on Computational Intelligence and Communication Networks (CICN)*, India, pp. 84–87. IEEE, 2019.

13. Pandey, Bishwajeet, and Ravikant Kumar. "Low voltage DCI based low power VLSI circuit implementation on FPGA." In *2013 IEEE Conference on Information & Communication Technologies*, India, pp. 128–131. IEEE, 2013.

14. Kaur, Harkinder, Harsh Sohal, and Jaiteg Singh. "Design and performance analysis of uart using Altera Quartus-II and Xilinx ISE 14.2." In *6th International Conference on Communication and Network Technologies*, India. 2016.

15. Kaur, Ravinder, Jagdish Kumar, Sumita Nagah, Bishwajeet Pandey, and Kavita Goswami. "IO Standard based low power memory design and implementation on FPGA." In *2015 2nd International Conference on Computing for Sustainable Global Development (INDIACom)*, India, pp. 1501–1505. IEEE, 2015.

16. Sharma, Rashmi, Bishwajeet Pandey, Vikas Jha, Siddharth Saurabh, and Sweety Dabas. "Input–output standard-based energy efficient UART design on 90nm FPGA." In Sunil Kumar Muttoo (Ed.), *System and architecture*, pp. 139–150. Springer, Singapore, 2018.

17. Kumar, Keshav, Amanpreet Kaur, and K. R. Ramkumar. "Effective data transmission with UART on Kintex-7 FPGA." In *2020 12th International Conference on Computational Intelligence and Communication Networks (CICN)*, India, pp. 492–497. IEEE, 2020.

18. Aggarwal, Arushi, Bishwajeet Pandey, Sweety Dabbas, Achal Agarwal, and Siddharth Saurabh. "LVCMOS-based low-power thermal-aware energy-proficient vedic multiplier design on different FPGAs." In Sunil Kumar Muttoo (Ed.), *System and architecture*, pp. 115–122. Springer, Singapore, 2018.
19. Kumar, Abhishek, Bishwajeet Pandey, D. M. Akbar Hussain, Mohammad Atiqur Rahman, Vishal Jain, and Ayoub Bahanasse. "Frequency scaling and high-speed transceiver logic based low power UART design on 45nm FPGA." In *2019 11th International Conference on Computational Intelligence and Communication Networks (CICN)*, India, pp. 88–92. IEEE, 2019.

Chapter 7

MOBILE DDR-based UART for GCC

LIST OF ABBREVIATIONS

ASIC	Application-Specific Integrated Circuit
BUFG	Global Buffers
Clk	Clocks
CMOS	Complementary Metal-Oxide Semiconductor
DP	Dynamic Power
DDR	Double Data Rate
FF	Flip-Flops
FPGA	Field Programmable Gate Array
GC	Green Computing
GCC	Green Communication Computing
G. Comm.	Green Communication
IC	Integrated Circuit
IT	Information Technology
IO	Input Output
JT	Junction Temperature
LP	Leakage Power
LPDDR	Low-Power Double Data Rate
LUT	Look-Up Tables
RTL	Register Transfer Logic
S/G	Signal
SP	Static Power
TM	Thermal Margin
TP	Total Power
TPC	Total Power Consumption
UART	Universal Asynchronous Receiver Transmitter
ϑJA	Effective Thermal Resistance to Air

DOI: 10.1201/9781003302872-7

7.1 INTRODUCTION

In the age of vast and rapid growth of technologies across the globe, we all are getting so much concise about the topic of power and energy resource management, because a balance is much required in the production and consumption of these resources. The term "energy and power deficiency" refers to a scenario in which the output of energy and power is insufficient to meet the demand [1]. As a result of this imbalance, the world is currently experiencing an energy crisis. An energy crisis is defined as any major stifling of an economy's availability of energy resources. The reasons of the energy problem are industrialization, scarcity, decreased output, overpopulation, and urbanization. Aside from these factors, there is one crucial factor to consider: overconsumption. Many sections of society use far more energy than is necessary. The global energy crisis is a major factor in the smooth operation of data transmission. When there is no power, the data communication process becomes stalled, resulting in data transmission delays. To accomplish proper communication, we need to shift to a system that uses less electricity [2,3]. To accomplish the same, in this chapter we interface the UART with Kintex-7 FPGA device, and to optimize the power consumption of the UART with the FPGA device, the impedance matching is being done with MOBILE DDR IO.

7.2 MOBILE DDR IO STANDARD

The MOBILE_DDR standard is for LPDDR (Low-Power Double Data Rate) and Mobile DDR memory buses. In the semiconductor industry, MOBILE DDR IO standards are used in FPGA and ASIC designs to match the impedance of input and output loads for optimizing the power. In this chapter, the impedance matching in UART design is implemented with MOBILE DDR IO for different frequencies ranging from 100 MHz to 5 GHz [4,5]. The implementation of MOBILE DDR IO at various frequency values helps in optimizing the power consumption of the device and thus helps in contributing toward the ideas of GC. The various frequency values at which MOBILE DDR IO is implemented are described in Figure 7.1.

Figure 7.1 Frequencies values at which MOBILE DDR IO is implemented.

7.3 INTEGRATION OF UART WITH FPGA

This section discusses about the RTL design, gate-level schematic circuit, and synthesized design of UART with Kintex-7 FPGA. The register transfer logic (RTL) design is described in Figure 7.2.

In Figure 7.2, there are two UARTs: one is for transmitting the signal, and the other is for receiving the signal. At the input of transmitting end, there are five wires (clk, 16-bit perscale, rst, s_axis_tdata 8 bit, and s_axix_tvalid), while at output side, there are three wires (busy, txd, and s_axix_tready). Similarly, at the input of receiving side there are six wires (clk, 16-bit perscale, rst, m_axis_tready, rst, and rxd) while at output side there are five wires (busy, frame_error, m_axis_tdata 8 bit, m_axis_tvalid, and overrun_error). The gate-level schematic circuit is illustrated in Figure 7.3, and the synthesized FPGA interfacing of UART is shown in Figure 7.4 [6–8].

7.4 RESOURCE UTILIZATION

On the Kintex-7 device, some FPGA resources are used in the implementation of UART. Look-up tables (LUTs), flip-flops (FF), input output (IO), and global buffers are some of the resources available (BUFG). Kintex-7 uses 79 FFs, 110 LUTs, 44 IOs, and 1 BUFG when implementing UART. Table 7.1 summarizes Kintex-7's resource usage, whereas Figure 7.5 depicts the percentage of resources consumed [9–11].

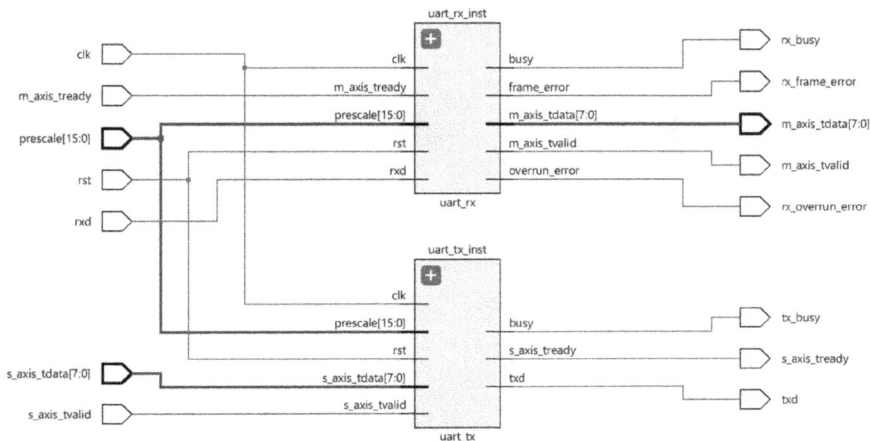

Figure 7.2 RTL design of UART.

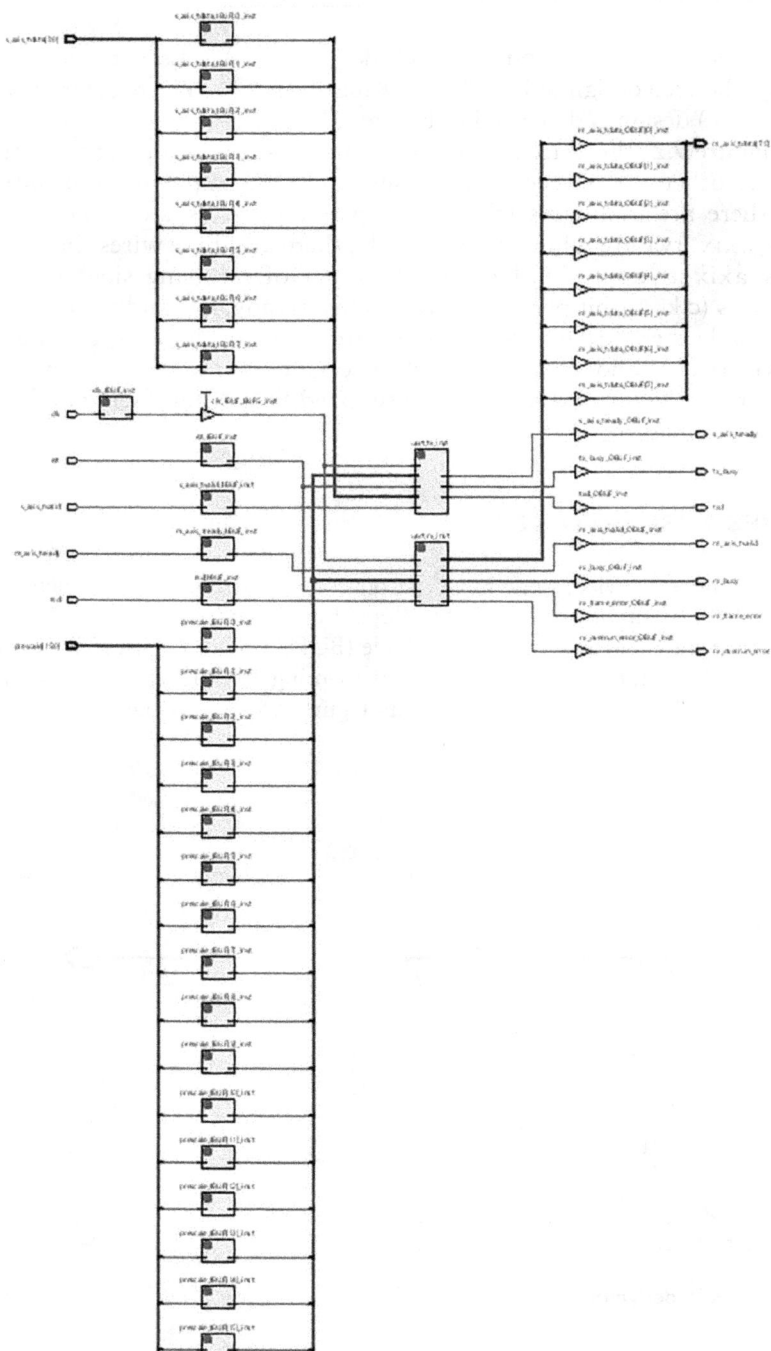

Figure 7.3 Gate-level schematic circuit.

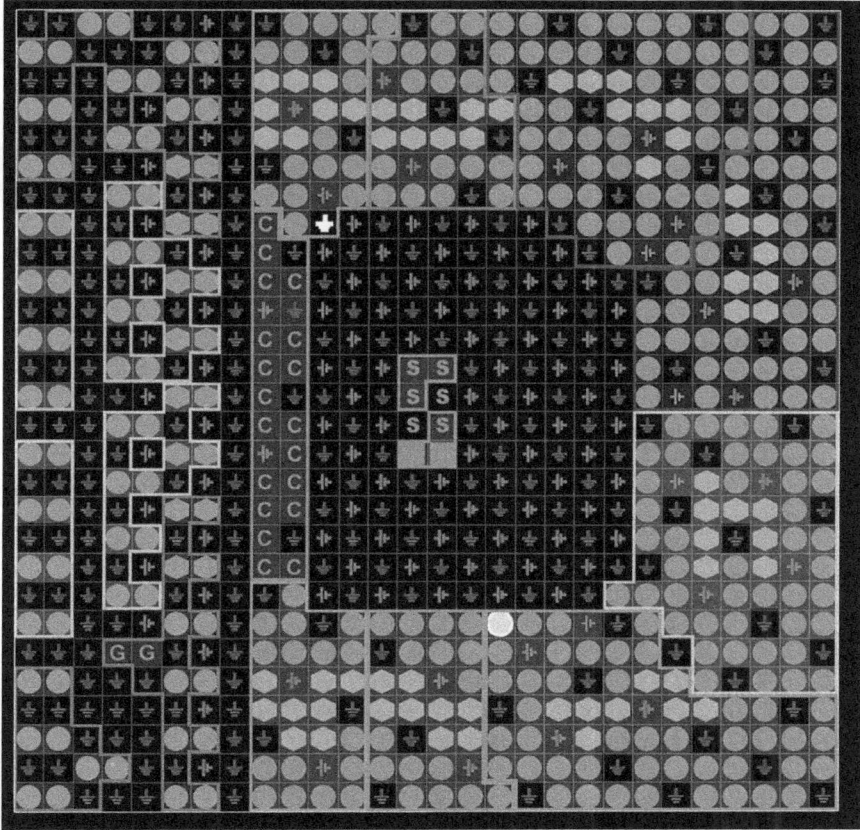

Figure 7.4 Synthesized FPGA interfacing of UART.

Table 7.1 Resource utilization of UART on Kintex-7

Resources	Utilization	Available	Utilization %
FF	79	43,3920	0.02
IO	44	304	14.47
LUT	110	21,6960	0.05
BUFG	1	256	0.39

Utilization %

BUFG	0.39
LUT	0.05
IO	14.47
FF	0.02

0   2   4   6   8   10   12   14   16

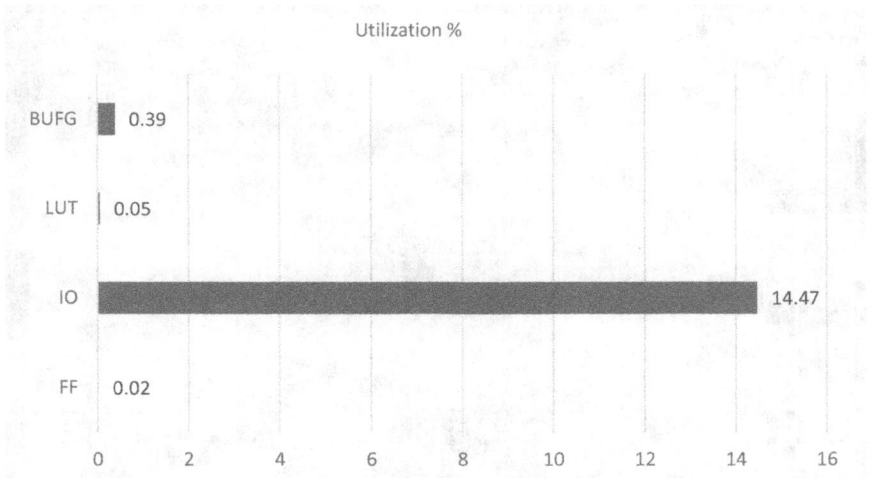

Figure 7.5 Percentage of resource consumed.

7.5 THERMAL PROPERTIES

This section gives insights about the Kintex-7 device's thermal properties when impedance matching is performed using MOBILE DDR IO standards for UART implementation on FPGA. Junction temperature (JT), thermal margin (TM), and effective thermal resistance (θJA) are the thermal parameters taken into consideration [12–14]:

a. JT: The highest operating temperature of an FPGA device's integrated gates.
b. TM: This is a device's attribute that permits it to consume very little power.
c. θJA: When 1 W of power is dissipated in the IC, the amount of heat created or a rise in temperature is specified.

Table 7.2 shows the thermal parameters of the MOBILE DDR IO standards for the different frequency values.

From Table 7.2, it is analyzed that for the frequency value between 100 MHz and 1 GHz, there is no change in any of the thermal properties for the MOBILE DDR IO. The change in thermal properties is observed only with JT and TM at 5 GHz frequency.

7.6 POWER ANALYSIS

The most devastating problem in the current scenario of the world is the power consumption in the electronics and IT technology. The whole planet

Table 7.2 Thermal properties associated with MOBILE DDR IO standards for Kintex-7

Frequency	JT (°C)	TM (°C)	ϑJA (°C/W)
100 MHz	25.3	74.7	1.9
300 MHz	25.3	74.7	1.9
500 MHz	25.3	74.7	1.9
1 GHz	25.3	74.7	1.9
5 GHz	25.5	74.5	1.9

is facing the problem of power deficiency. Thus, the whole ecosystem is collaborating to develop a low-power system for electronics and IT industry. Reduced power usage extends the device's life. In this section, we use the MOBILE DDR IO standard to calculate the power consumption of a UART implementation on a Kintex-7 device. The sum of the gadget's dynamic and static power consumption is its overall power consumption [15–17]. This is mathematically expressed in Equation 7.1.

$$TP = SP + DP \tag{7.1}$$

where

TP = Total Power
SP = Static Power
DP = Dynamic Power

The device SP is the total of the device's Clocks (clk), IO, Logic, and Signa (S/G) power, while the DP is the device's leakage power (LP).

7.6.1 Power analysis for 100 MHz frequency

When the frequency is tuned to 100 MHz value, the TPC of the device calculated is 0.137 W, which is the summation of DP (0.026 W) and SP (0.112 W). The TPC for 100 MHz is illustrated in Figure 7.6.

The DP is the total of clk, IO, logic, and S/G power, which are, respectively, 0.002, 0.023, <0.001, and <0.001 W. The SP is 0.112 W. The on-chips power for 100 MHz frequency is shown in Figure 7.7.

7.6.2 Power analysis for 300 MHz frequency

When the frequency is tuned to 300 MHz for matching the impedance of the device, the TP consumption of the device observed is 0.143 W. The device SP is 0.112 W which is 78% of TP, and DP is 0.031 W which is 22% of TP consumption. The power consumption of the UART for 300 MHz is illustrated in Figure 7.8. The graphical representation of TP consumption is described in Figure 7.9.

On-Chip Power

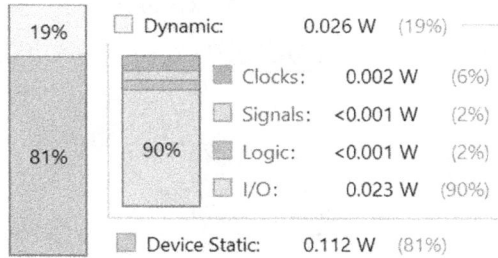

	Dynamic:	0.026 W	(19%)
	Clocks:	0.002 W	(6%)
	Signals:	<0.001 W	(2%)
	Logic:	<0.001 W	(2%)
	I/O:	0.023 W	(90%)
	Device Static:	0.112 W	(81%)

Figure 7.6 TPC for MOBILE DDR IO at 100MHz.

TPC (W) for 100 MHz

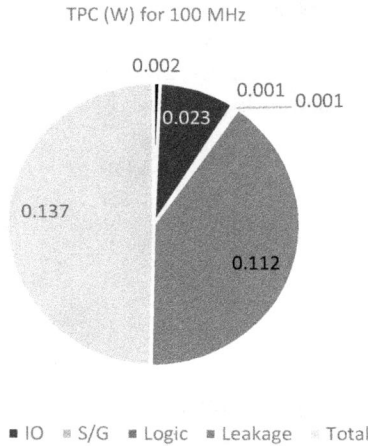

■ Clk   ■ IO   ▨ S/G   ■ Logic   ■ Leakage   ▨ Total

Figure 7.7 Representation of on-chips power for 100MHz.

On-Chip Power

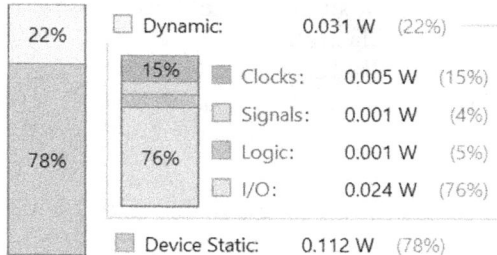

	Dynamic:	0.031 W	(22%)
	Clocks:	0.005 W	(15%)
	Signals:	0.001 W	(4%)
	Logic:	0.001 W	(5%)
	I/O:	0.024 W	(76%)
	Device Static:	0.112 W	(78%)

Figure 7.8 TPC for MOBILE DDR IO at 300MHz.

TPC (W) for 300 MHz

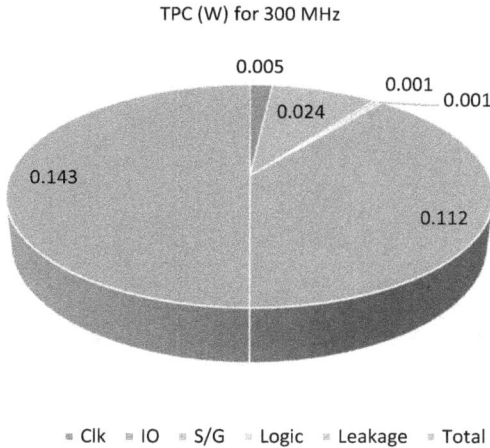

Figure 7.9 Representation of on-chips power for 300 MHz.

7.6.3 Power analysis for 500 MHz frequency

As the frequency of operation is increased to the value of 500 MHz, the TP of the device measured is 0.148 W, which is the cumulative sum of SP (0.112 W) which is 75% of the TP and DP (0.036 W); which is 25% of the TP. The DP is the total of clk, S/G, logic, and IO power, which are 0.008, 0.002, 0.002, and 0.024 W, respectively. The representation of the total on-chips power is shown in Figure 7.10, and the graphical representation is described in Figure 7.11.

7.6.4 Power analysis for 1 GHz frequency

When the frequency is tuned to 1 GHz for the impedance matching, it is observed that the total power consumption is 0.161 W, which is the summation of DP (0.050 W) and SP is (0.112 W). The TP consumption for 1 GHz is illustrated in Figure 7.12.

We know that the DP is the summation of clk, IO, S/G, and logic power. The clk, IO, S/G, and logic power are 0.016, 0.025, 0.004, and 0.005 W, respectively, for 1 GHz frequency. The leakage power is the SP which is 0.112 W. The graphical representation of on-chips power of the device is shown in Figure 7.13.

7.6.5 Power analysis for 5 GHz frequency

As the frequency of operation is increased to the value of 5 GHz, the TP of the device measured is 0.267 W, which is the cumulative sum of SP (0.112 W); which is 42% of the TP and DP (0.155 W) which is 58% of the TP. The DP

On-Chip Power

25%	☐ Dynamic: 0.036 W (25%)
	21% ▨ Clocks: 0.008 W (21%)
	☐ Signals: 0.002 W (6%)
75%	66% ▨ Logic: 0.002 W (7%)
	☐ I/O: 0.024 W (66%)
	▨ Device Static: 0.112 W (75%)

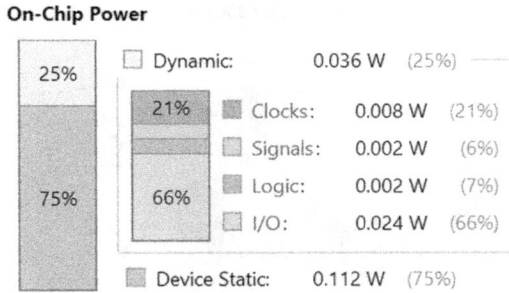

Figure 7.10 TPC of MOBILE DDR IO at 500 MHz frequency.

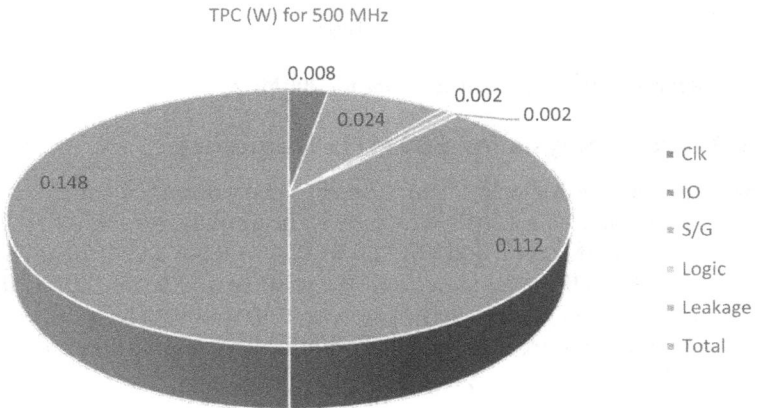

TPC (W) for 500 MHz

0.008
0.002
0.024
0.002
0.148
0.112

- Clk
- IO
- S/G
- Logic
- Leakage
- Total

Figure 7.11 Representation of on-chips power for 500 MHz.

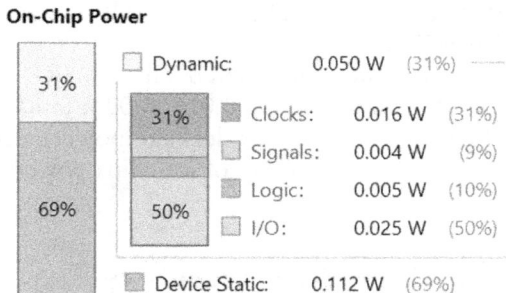

On-Chip Power

31%	☐ Dynamic: 0.050 W (31%)
	31% ▨ Clocks: 0.016 W (31%)
	☐ Signals: 0.004 W (9%)
69%	50% ▨ Logic: 0.005 W (10%)
	☐ I/O: 0.025 W (50%)
	▨ Device Static: 0.112 W (69%)

Figure 7.12 TP consumption for MOBILE DDR IO at 1 GHz frequency.

TPC (W) for 1GHz

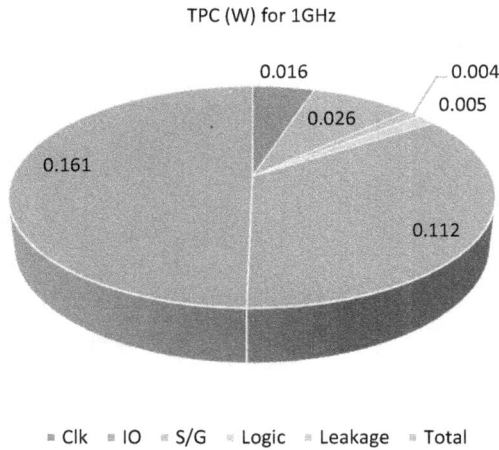

Clk IO S/G Logic Leakage Total

Figure 7.13 Representation of on-chips power for 1 GHz.

On-Chip Power

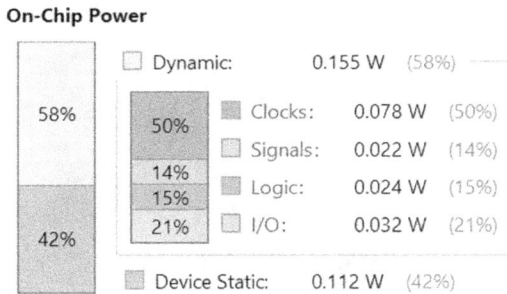

Dynamic:	0.155 W	(58%)
Clocks:	0.078 W	(50%)
Signals:	0.022 W	(14%)
Logic:	0.024 W	(15%)
I/O:	0.032 W	(21%)
Device Static:	0.112 W	(42%)

Figure 7.14 TP consumption for MOBILE DDR IO at 5 GHz frequency.

is the total of clk, S/G, logic, and IO power, which are 0.078, 0.022, 0.024, and 0.032 W respectively. The representation of the total on-chips power is shown in Figure 7.14, and the graphical representation is described in Figure 7.15.

7.7 OBSERVATION AND ANALYSIS

From the Section 7.6, it is observed that the TPC for the MOBILE DDR IO gets increased as the frequency of operation is increased for the device. The change in TP is observed because of the change in the DP of the FPGA device. The SP remains constant for the all-frequency range. This change

TPC (W) for 5 GHz

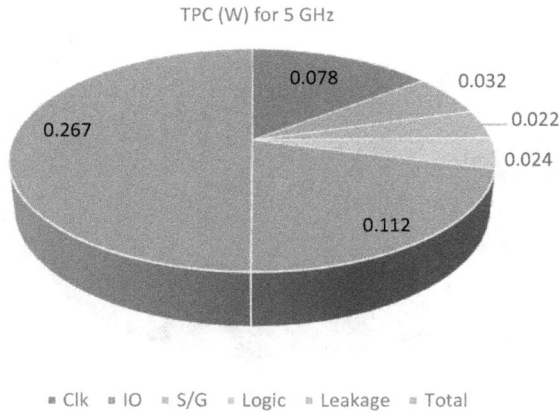

■ Clk   ■ IO   ■ S/G   ■ Logic   ■ Leakage   ■ Total

Figure 7.15 Representation of on-chips power for 5 GHz.

TPC (W)

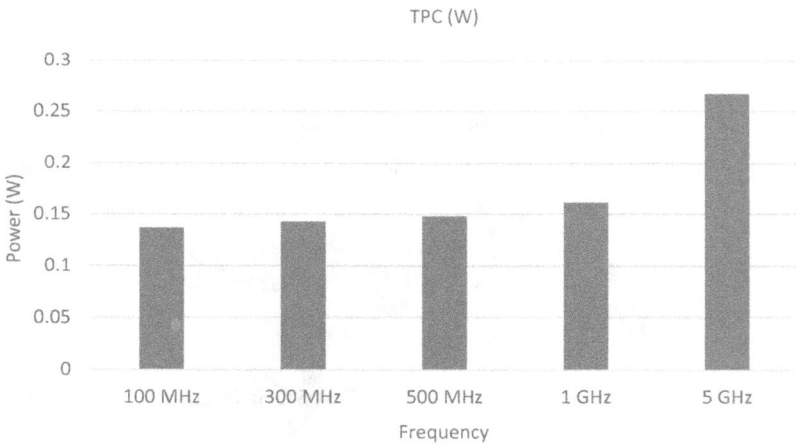

Figure 7.16 Variation in TPC for various frequency values.

in DP causes the increment in the TPC. The variation in TPC for various frequency values is described in Figure 7.16.

From Figure 7.16, it is observed that for MOBILE DDR IO, as the frequency of operation increases, the TPC of the device also increases. The device is optimal at 100 MHz frequency for the MOBILE DDR IO.

7.8 CONCLUSION

It has been observed that in past few years, the globe is facing huge problem of energy and power deficiency. The earth's natural resources may not live long. Someday it has to be vanished. In the context of such a huge problem the idea of GC comes in people's mind. This chapter is a step in the area of GCC. In this chapter, a power efficient model of UART is designed with the help of Kintex-7 device. To optimize the power utilization of UART, in this chapter, the authors have used the impedance matching technique, for which they have used the MOBILE DDR IO standards. In FPGA, IO standards are used to match the input and output impedance of the circuit. In this chapter, the TPC of the device is observed for a different set of frequency values for the MOBILE DDR IO. It is observed that the TPC of the device increases as the frequency of operation increases. The device gives optimal power for 100 MHz frequency value, and the device consumes the maximum power for the frequency of 5 GHz.

As far as the future scope is concerned, the UART design can be implemented with the other SoC-based FPGAs, and there are also different power optimization techniques that can be used such as capacitance scaling, voltage/current variation of the device, and clock gating. Also, the FPGA design can be converted into the ASIC design for better results.

REFERENCES

1. Gandotra, Pimmy, Rakesh Kumar Jha, and Sanjeev Jain. "Green communication in next generation cellular networks: A survey." *IEEE Access* 5 (2017): 11727–11758.
2. Mahapatra, Rajarshi, Yogesh Nijsure, Georges Kaddoum, Naveed Ul Hassan, and Chau Yuen. "Energy efficiency tradeoff mechanism towards wireless green communication: A survey." *IEEE Communications Surveys & Tutorials* 18, no. 1 (2015): 686–705.
3. Vereecken, Willem, Ward Van Heddeghem, Didier Colle, Mario Pickavet, and Piet Demeester. "Overall ICT footprint and green communication technologies." In *2010 4th International Symposium on Communications, Control and Signal Processing (ISCCSP)*, Limassol, Cyprus, pp. 1–6. IEEE, 2010.
4. Kumar, Tanesh, Bishwajeet Pandey, Teerath Das, and B. S. Chowdhry. "Mobile DDR IO standard based high performance energy efficient portable ALU design on FPGA." *Wireless Personal Communications* 76, no. 3 (2014): 569–578.
5. Agrawal, Tarun, Vivek Srivastava, and Anjan Kumar. "Designing of power efficient ROM using LVTTL and mobile-DDR IO standard on 28nm FPGA." In *2015 International Conference on Computational Intelligence and Communication Networks (CICN)*, India, pp. 1334–1337. IEEE, 2015.
6. Gupta, Isha, Swati Singh Garima, Harpreet Kaur, Deepshikha Bhatt, and Aamir Vohra. "28nm FPGA based power optimized UART design using HSTL I/O standards." *Indian Journal of Science and Technology* 8, no. 17 (2015): 1–6.

7. Sharma, Rashmi, Bishwajeet Pandey, Vikas Jha, Siddharth Saurabh, and Sweety Dabas. "Input–output standard-based energy efficient UART design on 90nm FPGA." In Sunil Kumar Muttoo (Ed.), *System and architecture*, pp. 139–150. Springer, Singapore, 2018.

8. Gupta, Isha, Swati Singh Garima, Harpreet Kaur, Deepshikha Bhatt, and Aamir Vohra. "28nm FPGA based power optimized UART design using HSTL I/O standards." *Indian Journal of Science and Technology* 8, no. 17 (2015): 1–6.

9. Kumar, Keshav, Amanpreet Kaur, S. N. Panda, and Bishwajeet Pandey. "Effect of different nano meter technology-based FPGA on energy efficient UART design." In *2018 8th International Conference on Communication Systems and Network Technologies (CSNT)*, India, pp. 1–4. IEEE, 2018.

10. Kumar, Keshav, Amanpreet Kaur, Bishwajeet Pandey, and S. N. Panda. "Low power UART design using different nanometer technology based FPGA." In *2018 8th International Conference on Communication Systems and Network Technologies (CSNT)*, India, pp. 1–3. IEEE, 2018.

11. Kumar, Keshav, Bishwajeet Pandey, Amit Kant Pandit, Y. A. Baker El-Ebiary, Salameh A. Mjlae, and Samer Bamansoor. "Design of low power transceiver on Spartan-3 and Spartan-6 FPGA." *International Journal of Innovative Technology and Exploring Engineering* 8, no. 12S2 (2019): 27–30.

12. Sandhu, Amanpreet, Vidhoytma Gandhi, Simranpreet Kaur, Surbhi Huria, Divjot Singh, and Wamika Goyal. "Thermally aware LVCMOS based low power universal asynchronous receiver transmitter design on FPGA." *Indian Journal of Science and Technology* 8, no. 20 (2015): 1–4.

13. Kumar, Abhishek, Bishwajeet Pandey, D. M. Akbar Hussain, Mohammad Atiqur Rahman, Vishal Jain, and Ayoub Bahanasse. "Low voltage complementary metal oxide semiconductor based energy efficient UART design on Spartan-6 FPGA." In *2019 11th International Conference on Computational Intelligence and Communication Networks (CICN)*, India, pp. 84–87. IEEE, 2019.

14. Pandey, Bishwajeet, and Ravikant Kumar. "Low voltage DCI based low power VLSI circuit implementation on FPGA." In *2013 IEEE Conference on Information & Communication Technologies,* India, pp. 128–131. IEEE, 2013.

15. Kaur, Harkinder, Harsh Sohal, and Jaiteg Singh. "Design and performance analysis of UART using Altera Quartus-II and Xilinx ISE 14.2." In *6th International Conference on Communication and Network Technologies.* 2016.

16. Kaur, Ravinder, Jagdish Kumar, Sumita Nagah, Bishwajeet Pandey, and Kavita Goswami. "IO Standard based low power memory design and implementation on FPGA." In *2015 2nd International Conference on Computing for Sustainable Global Development (INDIACom)*, India, pp. 1501–1505. IEEE, 2015.

17. Sharma, Rashmi, Bishwajeet Pandey, Vikas Jha, Siddharth Saurabh, and Sweety Dabas. "Input–output standard-based energy efficient UART design on 90nm FPGA." In Sunil Kumar Muttoo (Ed.), *System and architecture*, pp. 139–150. Springer, Singapore, 2018.

Chapter 8

LVCMOS-based FIR filter for GCC

LIST OF ABBREVIATIONS

AC	Alternating Current
ASIC	Application-Specific Integrated Circuit
BUFG	Global Buffers
Clk	Clocks
CMOS	Complementary Metal-Oxide Semiconductor
DP	Dynamic Power
FF	Flip-Flops
FIR Filter	Finite Impulse Response Filter
FPGA	Field Programmable Gate Array
GC	Green Computing
GCC	Green Communication Computing
G Comm.	Green Communication
IC	Integrated Circuit
IIR Filter	Infinite Impulse Response Filter
IT	Information Technology
IO	Input Output
JT	Junction Temperature
LP	Leakage Power
LVCMOS	Low-Voltage CMOS: Complementary Metal-Oxide Semiconductor
LUT	Look Up Tables
MOSFET	Metal-Oxide Semiconductor Field-Effect Transistor
RTL	Register Transfer Logic
SD	Sustainable Development
S/G	Signal
SP	Static Power
TM	Thermal Margin
TP	Total Power
TPC	Total Power Consumption
ϑJA	Effective Thermal Resistance to Air

DOI: 10.1201/9781003302872-8

8.1 INTRODUCTION

In digital signal processing, a FIR is a filter with a finite-period impulse response that settles to zero in limited time. This differs from IIR filters, which may have internal feedback and continue to react forever. A Nth-order, discrete-time FIR filter's impulse response takes exactly N+1 sample before settling to zero. FIR filters are the most common kind of software-based filters, and they may be continuous time, analog or digital, or discrete time. Boxcar, Hilbert Transformer, Differentiator, Lth-Band, and Raised-Cosine are examples of special kinds of FIR filters [1,2].

The phrase "Finite Impulse Response" is one of the two primary kinds of digital filters used in DSP applications. Filters are signal conditioners, and their role is to allow AC components while blocking DC components. A phone line is the finest illustration of a filter, because it restricts the spectrum of frequencies that humans can hear [3,4].

8.1.1 Explanation

The FIR filter for a causal time signal is expressed as follows:

$$a[n] = b_0 c[n] + b_1 c[n-1] + ... + b_n c[n-N]$$

$$= \sum_{i=0}^{N} b_i \bullet c[n-i]$$

where

- $c[n]$ is the input signal
- $a[n]$ is the output signal
- N is the order of filter
- b_i is the value of the impulse response for FIR filter

8.1.2 Properties

There are a number of handy properties of FIR filter which makes it preferable to use in comparison to an IIR (infinite impulse response) filter such as:

a. No feedback is required in the signal processing of the FIR filter
b. These filters are more stable as compared to IIR filter
c. The design of FIR filter is also easier
d. These filters are phase sensitive
e. Hardware implementation of FIR filter is easier

Therefore, we all know that there are numerous usages of FIR filter in signal processing and digital technology. These filters consume huge amount of power in signal processing [5]. The consumption of huge power requires a big battery in system, or it requires more usage of natural resources for power consumption. The power deficiency can be observed in the whole globe. In order to keep such things in our minds, a low-power system must be designed so that the system life can be increased more and also be beneficial for GC [6,7]. The GC design of FIR filter can be very handy in promoting the sustainable development (SD) across the globe. Hence, in this chapter, we design such a system that will promote the ideas of SD as well as GC. For such a system, we use SPARTAN-7 FPGA and the power consumption is optimized using the LVCMOS IO standard.

8.2 FPGA IMPLEMENTATION OF FIR FILTER

To design a power-efficient model of FIR filter, we have used the SPARTAN-7 FPGA device. It is a 28 nm gate sized device of Xilinx family. The FIR design is implemented on the same device using the VIVADO ISE. To optimize the power utilization, LVCMOS IOs are used. IO standard matches the transmission line impedance for the internal circuit so that the power utilization should be minimized. In order to implement the design on the FPGA device, FPGA resources are consumed such as IO, FF, LUTs, and BUFG. The resource utilization for designing FIR design is shown in Table 8.1 and described in Figure 8.1 [8–10].

From Table 8.1, it is observed that in the implementation process the FIR design 19 IO and LUTs are consumed as well as 28 FFs and 1 BUFG of SPARTAN-7 device. The RTL schematic for the FIR filter is shown in Figure 8.2, while the synthesized RTL design of the FIR is depicted in Figure 8.3.

Table 8.1 Resource utilization for designing FIR design

Resources	Available	Utilization	Utilization%
LUT	19	48,000	0.04
IO	19	338	5.62
BUFG	1	32	3.13
FF	28	96,000	0.03

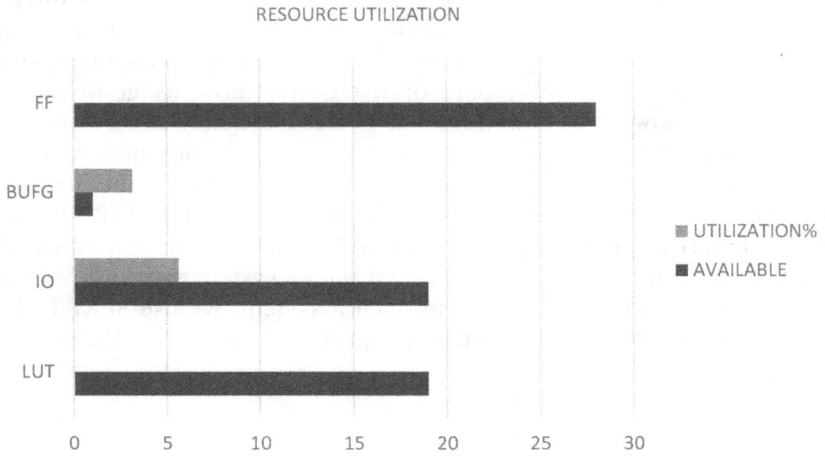

Figure 8.1 Resource utilization for designing FIR design.

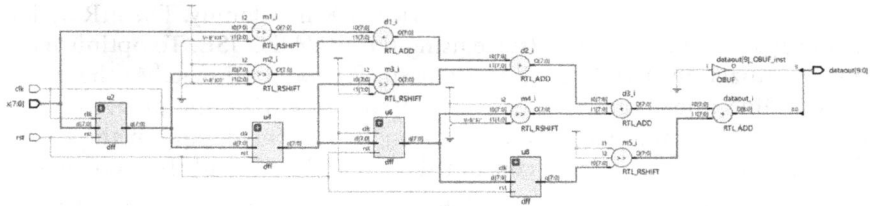

Figure 8.2 RTL schematic of FIR filter.

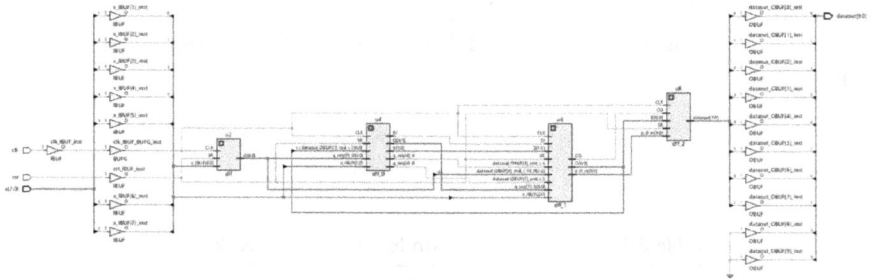

Figure 8.3 Synthesized RTL design of FIR filter.

8.3 THERMAL PROPERTIES

This section will give a brief insight about the thermal properties of the device. The thermal properties are taken into account while matching the impedance using the LVCMOS IO standard. The thermal properties considered here are as follows:

a. TM: This is a device's attribute that permits it to consume very little power.
b. ϑJA: When 1 W of power is dissipated in the IC, the amount of heat created or a rise in temperature is specified.
c. JT: The highest operating temperature of an FPGA device's integrated gates.

The thermal properties for the LVCMOS IO for SPRATAN-7 are described in Table 8.2. From Table 8.2, it observed that the JT increases as the input voltage of the transmission line impedance matching increases. The TM decreases as the input voltage increases LVCMOS IO. For the LVCMOS 33 IO, the TM value becomes negative. Also, for the LVCMOS 33, the junction gate of the internal-connected MOSFET burns (red color). The ϑJA value remains constant to 2.7 for all LVCMOS IO. The variation in thermal properties is described in Figure 8.4 [11–13].

Table 8.2 Thermal properties for the LVCMOS IO for SPRATAN-7

IO std	JT (°C)	TM (°C)	ϑJA (°C/W)
LVCMOS 12	26.4	58.6	2.7
LVCMOS 15	41.1	43.9	2.7
LVCMOS 18	46.2	38.8	2.7
LVCMOS 25	61.5	23.5	2.7
LVCMOS 33	87.7	−2.2	2.7

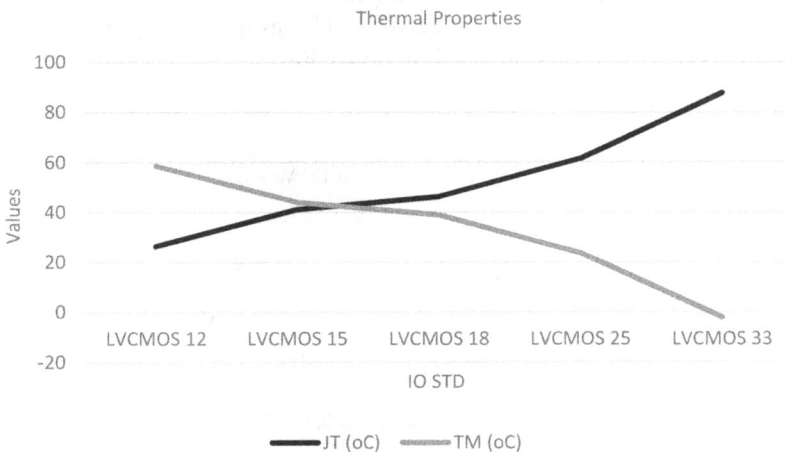

Figure 8.4 Thermal properties for the LVCMOS IO for SPRATAN-7.

8.4 POWER ANALYSIS

In the present state of the globe, the power consumption of gadgets and information technology poses the greatest threat. The whole globe is experiencing an energy deficit. Thus, the whole ecosystem collaborates to produce a low-power solution for the electronics and information technology industries. Reduced power consumption prolongs the device's life. In this part, we calculate the power consumption of a FIR implementation on a SPARTAN-7 device using the LVCMOS IO standard. The device's total power consumption is the sum of both dynamic and static power consumption. This is described mathematically in Equation 8.1 [14–16]:

$$TP = SP + DP \qquad (8.1)$$

where

TP = Total Power
SP = Static Power
DP = Dynamic Power

8.4.1 Power analysis for LVCMOS 12

When the impedance is matched with the LVCMOS IO, the total power consumption (TPC) observed for the device is 0.523W. The DPC is 83% of the TPC which is 0.433 W, and SPC is 17% of the TPC which is 0.090 W. The on-chips power consumption for the device when impedance is matched with LVCMOS 12 IO is described in Figure 8.5.

From Figure 8.5, it can be clearly observed that the DPC is the combination of signal (S/G), logic (L/G), and IO power of the device. These are 0.277, 0.152, and 0.004 W, respectively. The graphical representation of TPC for LVCMOS 12 is described in Figure 8.6.

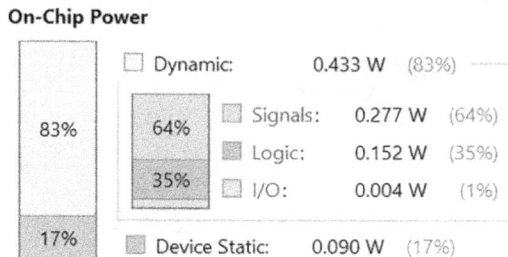

Figure 8.5 On-chips power consumption for LVCMOS 12 IO.

TPC (W) for LVCMOS 12 IO

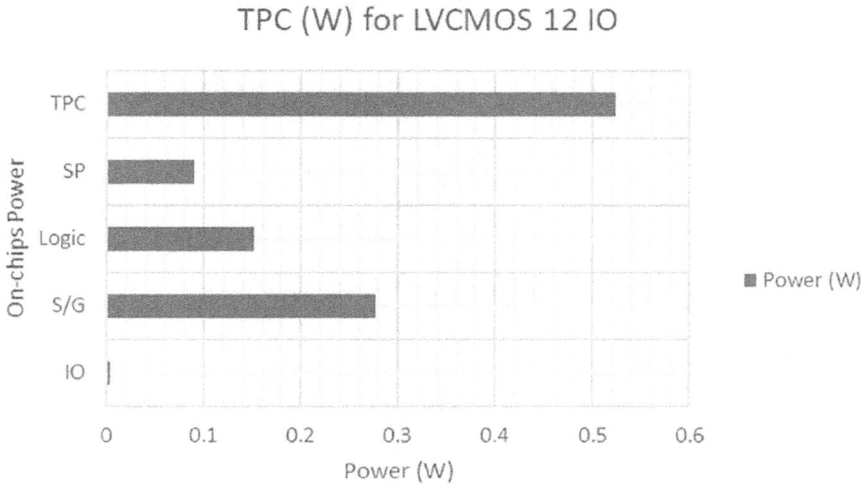

Figure 8.6 Graphical representation of the TPC for LVCMOS 12.

8.4.2 Power analysis for **LVCMOS 15**

When the LVCMOS 15 IO is used for matching the impedance of the transmission line, it is observed that the DPC is 5.783 W, which is 98% of the TPC of the device. The SPC is just 2% of the TPC, which is 0.118 W. The TPC is the summation of SPC and DPC which is 5.901 W. The on-chips power consumption is described in Figure 8.7.

The DPC is the combination of S/G, L/G, and IO power of the device which are 0.276, 0.152, 5.354 W, respectively. The graphical representation of the TPC for LVCMOS 15 is presented in Figure 8.8.

8.4.3 Power analysis for **LVCMOS 18**

When the impedance is matched with the LVCMOS 18 IO, the total power consumption (TPC) observed for the device is 7.784 W. The DPC is 98% of the TPC which is 7.658 W, and SPC is 2% of the TPC which is 0.126 W. The on-chips power consumption for the device when impedance is matched with LVCMOS 18 IO is described in Figure 8.9.

From Figure 8.9, it can be clearly observed that the DPC is the combination of signal (S/G), logic (L/G), and IO power of the device. These are 0.276, 0.152, and 7.230 W, respectively. The graphical representation of the TPC for LVCMOS 18 is described in Figure 8.10.

On-Chip Power

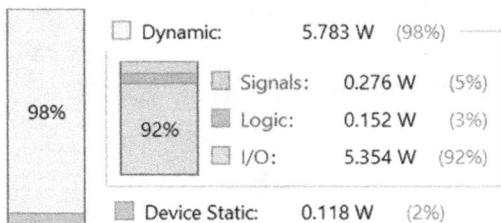

Dynamic:	5.783 W	(98%)
Signals:	0.276 W	(5%)
Logic:	0.152 W	(3%)
I/O:	5.354 W	(92%)
Device Static:	0.118 W	(2%)

Figure 8.7 On-chips power consumption for LVCMOS 15.

TPC (W) for LVCMOS 15 IO

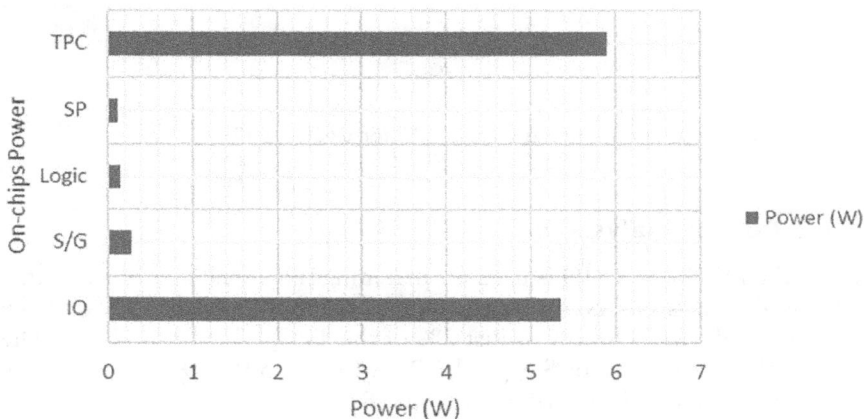

Figure 8.8 Graphical representation of the TPC for LVCMOS 15.

On-Chip Power

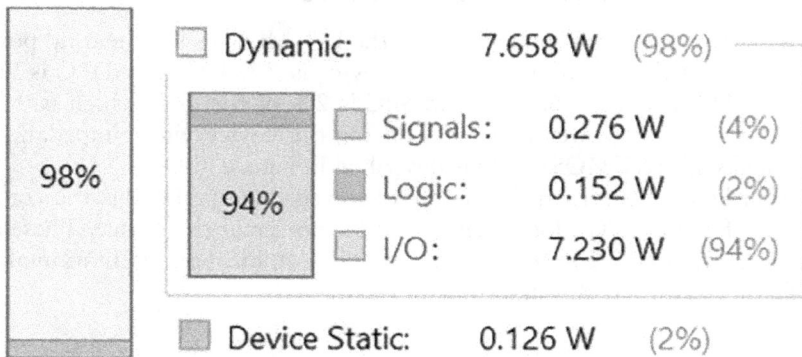

Dynamic:	7.658 W	(98%)
Signals:	0.276 W	(4%)
Logic:	0.152 W	(2%)
I/O:	7.230 W	(94%)
Device Static:	0.126 W	(2%)

Figure 8.9 On-chips power consumption for LVCMOS 18 IO.

TPC (W) for LVCMOS 18 IO

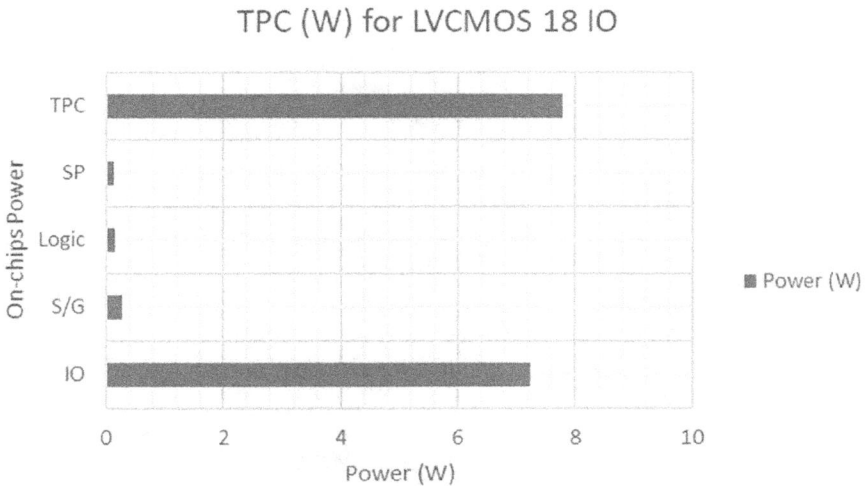

Figure 8.10 Graphical representation of the TPC for LVCMOS 18.

8.4.4 Power analysis for **LVCMOS 25**

When the LVCMOS 25 IO is used for matching the impedance of the transmission line, it is observed that the DPC is 13.256 W, which is 99% of the TPC of the device. The SPC is just 1% of the TPC, which is 0.173 W. The TPC is the summation of SPC and DPC which is 13.429 W. The on-chips power consumption is described in Figure 8.11.

The DPC is the combination of S/G, L/G, and IO power of the device which are 0.277, 0.152, 12.826 W, respectively. The graphical representation of the TPC for LVCMOS 25 is presented in Figure 8.12.

8.4.5 Power analysis for **LVCMOS 33**

When the impedance is matched with LVCMOS 33 IO, it is observed that the JT of the device gets exceeded and hence the junction gate of the device gets burned. The TPC of the devices calculated is 22.867 W. The on-chips power consumption of the device is represented in Figure 8.13.

8.5 OBSERVATION AND ANALYSIS

This section will give insights about the analysis and observation from Section 8.4. From Section 8.4, it is observed that the TPC increases as the input voltage of the transmission line increases; i.e., TPC increases for LVCMOS IO are shown in Figure 8.14.

On-Chip Power

	Dynamic: 13.256 W (99%)
99%	Signals: 0.277 W (2%)
97%	Logic: 0.152 W (1%)
	I/O: 12.826 W (97%)
	Device Static: 0.173 W (1%)

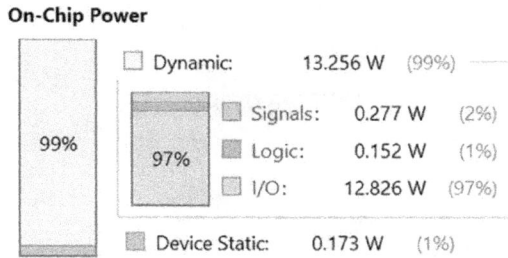

Figure 8.11 On-chips power consumption for LVCMOS 25.

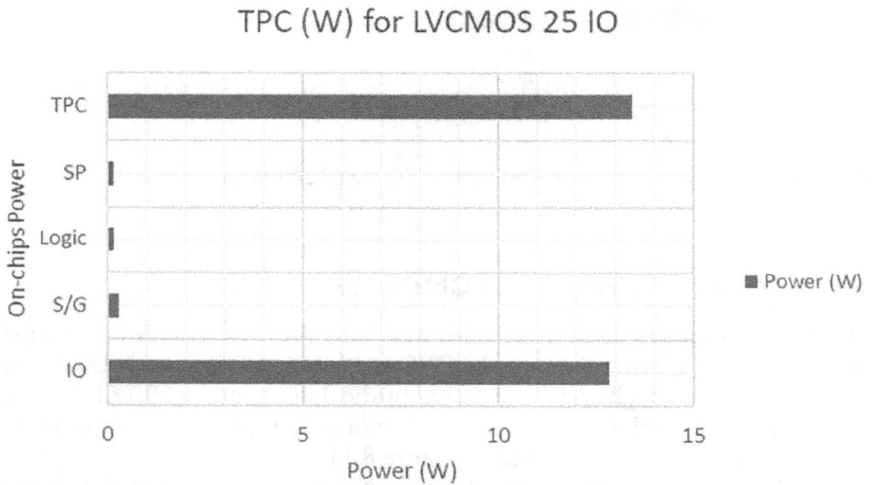

TPC (W) for LVCMOS 25 IO

Figure 8.12 Graphical representation of the TPC for LVCMOS 25.

Power analysis from Implemented netlist. Activity derived from constraints files, simulation files or vectorless analysis.

Total On-Chip Power:	22.867 W (Junction temp exceeded!)
Design Power Budget:	Not Specified
Power Budget Margin:	N/A
Junction Temperature:	87.2°C
Thermal Margin:	-2.2°C (-0.7 W)
Effective ϑJA:	2.7°C/W
Power supplied to off-chip devices:	0 W
Confidence level:	Low

Launch Power Constraint Advisor to find and fix invalid switching activity

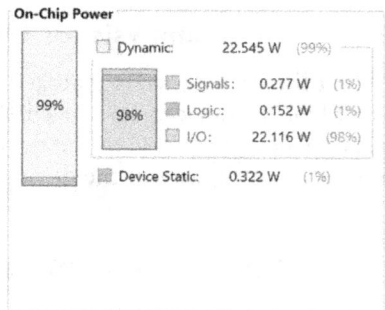

On-Chip Power

	Dynamic: 22.545 W (99%)
99%	Signals: 0.277 W (1%)
98%	Logic: 0.152 W (1%)
	I/O: 22.116 W (98%)
	Device Static: 0.322 W (1%)

Figure 8.13 On-chips power for LVCMOS 33 IO.

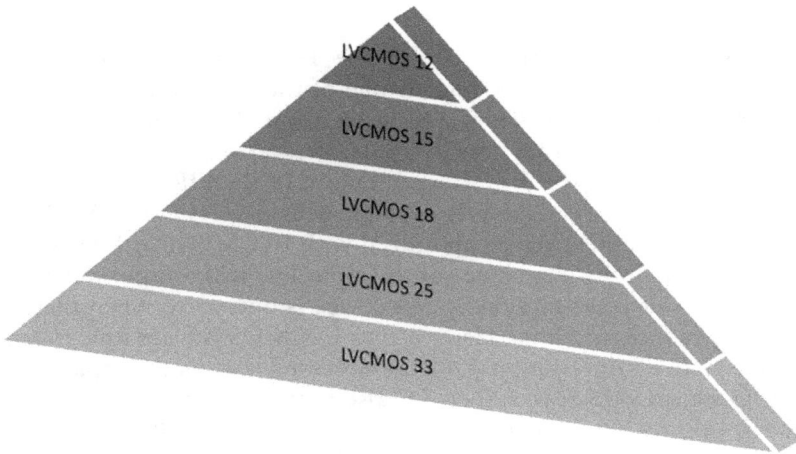

Figure 8.14 TPC increasing for LVCMOS IO.

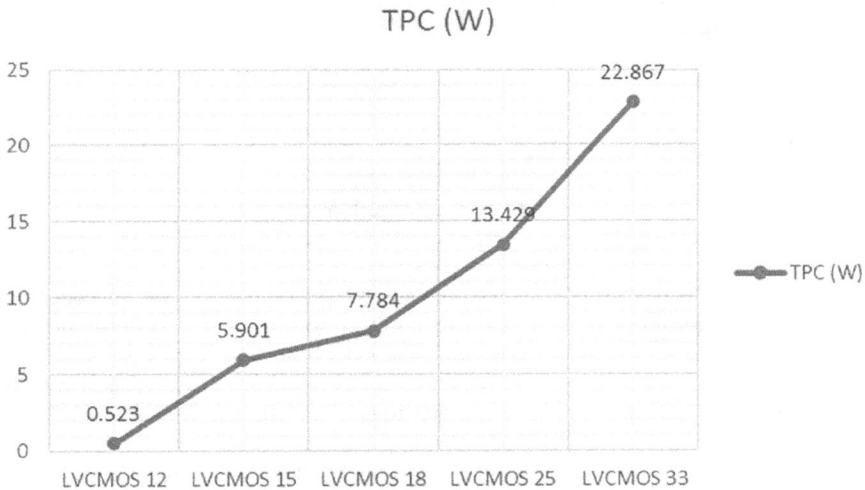

Figure 8.15 TPC of the device for LVCMOS family.

From Section 8.4, it is observed that TPC is minimum for LVCMOS 12 IO, and the TPC is maximum for LVCMOS 33 IO. Also, when the imped-ance is matched with LVCMOS 33 IO, the JT gets exceeded and hence the junction gate gets burned, so the device fails to work. The TPC of the device for LVCMOS family IO is described in Figure 8.15.

8.6 CONCLUSION

This chapter gives a detailed view of the TPC of the FIR filter on the SPARTAN-7 FPGA. The FIR filter is one of the key components used in the world of signal processing, and apart from this, it has also several DSP applications. The FIR filter has lots of several advantages over FIR filter. In this chapter, the power consumption of the FIR is analyzed over SPARTAN-7. The design of FIR is implemented on VIVADO ISE. The power consumption is tried to optimized with LVCMOS IO. IO standards are used to match the impedance of the input line with output line impedance. The power consumption of any device is optimized when there is a perfect impedance matching. Here various LVCMOS IOs are used to match the impedance. From Section 8.4, it is observed that as the input voltage of the IO standard increases, the TPC also increases. Therefore, the device gives the optimal power consumption when the input voltage is low. The FIR design on SPARTAN-7 devices has the optimal power consumption when the impedance is matched with LVCMOS 12 IO. As far as future scope is concerned, the FIR design can be implemented over several other FPGAs of distinguished family such as Artix-7, Kinten-7, and Virtex-7. Also, the designs can be converted into ASIC design for better applications and uses.

REFERENCES

1. Koilpillai, R. David, and P. P. Vaidyanathan. "Cosine-modulated FIR filter banks satisfying perfect reconstruction." *IEEE Transactions on Signal Processing* 40, no. 4 (1992): 770–783.
2. Vaidyanathan, P., and Truong Nguyen. "Eigenfilters: A new approach to least-squares FIR filter design and applications including Nyquist filters." *IEEE Transactions on Circuits and Systems* 34, no. 1 (1987): 11–23.
3. Lim, Yong, and Sydney Parker. "FIR filter design over a discrete powers-of-two coefficient space." *IEEE Transactions on Acoustics, Speech, and Signal Processing* 31, no. 3 (1983): 583–591.
4. Hsiao, Shen-Fu, Jun-Hong Zhang Jian, and Ming-Chih Chen. "Low-cost FIR filter designs based on faithfully rounded truncated multiple constant multiplication/accumulation." *IEEE Transactions on Circuits and Systems II: Express Briefs* 60, no. 5 (2013): 287–291.
5. Vetterli, Martin, and Didier Le Gall. "Perfect reconstruction FIR filter banks: Some properties and factorizations." *IEEE Transactions on Acoustics, Speech, and Signal Processing* 37, no. 7 (1989): 1057–1071.
6. Mahapatra, Rajarshi, Yogesh Nijsure, Georges Kaddoum, Naveed Ul Hassan, and Chau Yuen. "Energy efficiency tradeoff mechanism towards wireless green communication: A survey." *IEEE Communications Surveys & Tutorials* 18, no. 1 (2015): 686–705.

7. Vereecken, Willem, Ward Van Heddeghem, Didier Colle, Mario Pickavet, and Piet Demeester. "Overall ICT footprint and green communication technologies." In *2010 4th International Symposium on Communications, Control and Signal Processing (ISCCSP),* Limassol, Cyprus, pp. 1–6. IEEE, 2010.

8. Kumar, Tanesh, Bishwajeet Pandey, Teerath Das, and B. S. Chowdhry. "Mobile DDR IO standard based high performance energy efficient portable ALU design on FPGA." *Wireless Personal Communications* 76, no. 3 (2014): 569–578.

9. Agrawal, Tarun, Vivek Srivastava, and Anjan Kumar. "Designing of power efficient ROM using LVTTL and mobile-DDR IO standard on 28nm FPGA." In *2015 International Conference on Computational Intelligence and Communication Networks (CICN),* India, pp. 1334–1337. IEEE, 2015.

10. Gupta, Isha, Swati Singh Garima, Harpreet Kaur, Deepshikha Bhatt, and Aamir Vohra. "28nm FPGA based power optimized UART design using HSTL I/O standards." *Indian Journal of Science and Technology* 8, no. 17 (2015): 1–6.

11. Sharma, Rashmi, Bishwajeet Pandey, Vikas Jha, Siddharth Saurabh, and Sweety Dabas. "Input–output standard-based energy efficient UART Design on 90nm FPGA." In Sunil Kumar Muttoo (Ed.), *System and architecture,* pp. 139–150. Springer, Singapore, 2018.

12. Gupta, Isha, Swati Singh Garima, Harpreet Kaur, Deepshikha Bhatt, and Aamir Vohra. "28nm FPGA based power optimized UART design using HSTL I/O standards." *Indian Journal of Science and Technology* 8, no. 17 (2015): 1–6.

13. Kumar, Keshav, Amanpreet Kaur, S. N. Panda, and Bishwajeet Pandey. "Effect of different nano meter technology-based FPGA on energy efficient UART design." In *2018 8th International Conference on Communication Systems and Network Technologies (CSNT),* India, pp. 1–4. IEEE, 2018.

14. Kumar, Keshav, Amanpreet Kaur, Bishwajeet Pandey, and S. N. Panda. "Low power UART design using different nanometer technology based FPGA." In *2018 8th International Conference on Communication Systems and Network Technologies (CSNT),* India, pp. 1–3. IEEE, 2018.

15. Pandey, Bishwajeet, and Ravikant Kumar. "Low voltage DCI based low power VLSI circuit implementation on FPGA." In *2013 IEEE Conference on Information & Communication Technologies,* India, pp. 128–131. IEEE, 2013.

16. Kaur, Harkinder, Harsh Sohal, and Jaiteg Singh. "Design and performance analysis of uart using Altera Quartus-II and Xilinx ISE 14.2." In *6th International Conference on Communication and Network Technologies,* India. 2016.

Chapter 9

SSTL-based FIR filter of GCC

LIST OF ABBREVIATIONS

AC	Alternating Current
ASIC	Application-Specific Integrated Circuit
BUFG	Global Buffers
Clk	Clocks
DDR	Double Data Rate
DRAM	Dynamic Random-Access Memory
DP	Dynamic Power
FF	Flip-Flops
FPGA	Field Programmable Gate Array
GC	Green Computing
GCC	Green Communication Computing
G Comm.	Green Communication
GPS	Global Positioning System
IC	Integrated Circuit
IO	Input Output
JT	Junction Temperature
LP	Leakage Power
LUT	Look Up Tables
RTL	Register Transfer Logic
SSTL	Stub Series Terminated Logic
S/G	Signal
SP	Static Power
TM	Thermal Margin
TP	Total Power
TPC	Total Power Consumption
UART	Universal Asynchronous Receiver Transmitter
ϑJA	Effective Thermal Resistance to Air

DOI: 10.1201/9781003302872-9

9.1 INTRODUCTION

In a variety of modern applications using integrated circuits, increasing speed and decreasing power consumption are the essential objectives. Using cutting-edge complementary metal-oxide-semiconductor (CMOS) technology, these specifications are reached owing to the lowering device sizes and low voltage supply. Numerous previous analog circuit architectures, such as analog-to-digital converters, struggle to overcome the lower voltage headroom and narrower dynamic range caused by smaller device sizes and low-power supply. The time-domain approach is a relatively recent method of processing time that uses time delay, time difference, or pulse width instead of voltage or current, as is the case with standard processing approaches [1,2]. Therefore, the important quantity in time-domain circuits and systems is time. Even with such a little technology, time-domain design is a very promising design method since it offers a wider dynamic range to power consumption trade-off [3]. The advantage of time-domain systems is that they use high-speed MOS transistors, meaning that they have a shorter time delay and can thus manage time more precisely. The primary advantages of time-domain design are enhanced dynamic range and time resolution in comparison to analog voltage or current mode circuits in the same low-supply environment, as well as improved power efficiency for high-speed performance due to their composition of primarily CMOS digital building blocks (gates, etc.). FIR implementations need z-1 operators, signal adders, and signal multipliers for filter coefficient creation [4,5]. Traditional FIR implementations are discrete-time/discrete-signal processing systems largely based on a digital design technique. The basic operators are implemented as circuits employing flip-flops as delay components and digital logic gates to create digital logic adders and multipliers. The FIR filter has several numerous advantages in electronics and communication filed. The advancement in technologies has also allowed the FIR filter to be used in other sectors. The numerous applications use huge amount of power, and this huge consumption of power makes the circuitry bulky since the battery size might be increased. Apart from this, the lifetime may also be decreased. Therefore, the FIR implementation is to be done in such a way that it should consume low power. This chapter focuses on the low-power implementation of FIR filter [6–8]. To make the design power-efficient, the FIR design is implemented on a FPGA device. The FPGA devices have abundant applications in electronics and communication technologies. Also, the FPGA devices help in optimizing the power consumption of the implemented circuit. To make the FIR design power-efficient, Zynq-7000 FPGA device is used with SSTL IO standard.

Zynq-7000 devices include dual-core ARM Cortex-A9 processors combined with 28 nm Artix-7 or Kintex-7 programmable logic for superior performance-per-watt and maximum design freedom. Zynq-7000 devices allow highly distinctive designs for a broad variety of embedded applications, such as multi-camera driver assistance systems and 4K2K Ultra-HDTV, with up to 6.6M logic cells and transceivers ranging from 6.25 to 12.5 Gb/s. Zynq-7000 delivers power-optimized system integration suitable for IoT applications, communication channels, and embedded programs [9,10].

9.2 FIR FILTER IMPLEMENTATION ON ZYNQ-7000 FPGA DEVICE

To design a power-efficient model of FIR filter, we have used the Zynq-7000 FPGA device. The FIR design is implemented on the same device using the VIVADO ISE. To optimize the power utilization, SSTL IOs are used. IO standard matches the transmission line impedance for the internal circuit so that the power utilization should be minimized. In order to implement the design on the FPGA device, FPGA resources are consumed such as IO, FF, LUTs, and BUFG. The resource utilization for designing FIR design is shown in Table 9.1 and described in Figure 9.1 [11,12].

Table 9.1 Resource utilization for designing FIR design

Resources	Available	Utilization	Utilization%
LUT	19	53,200	0.04
IO	19	200	9.50
BUFG	1	32	3.13
FF	28	10,6400	0.03

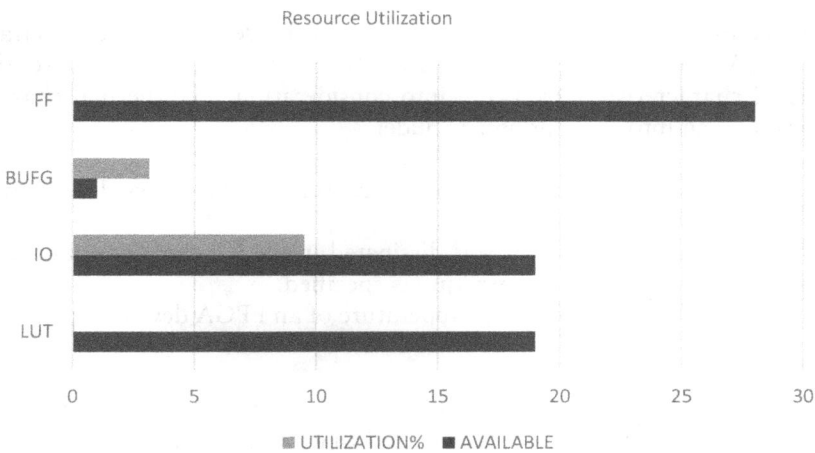

Figure 9.1 Resource utilization for designing FIR design.

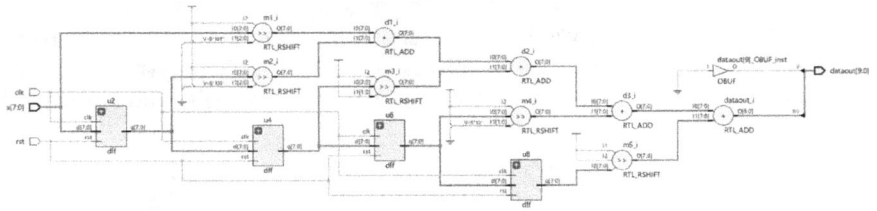

Figure 9.2 RTL schematic of FIR filter.

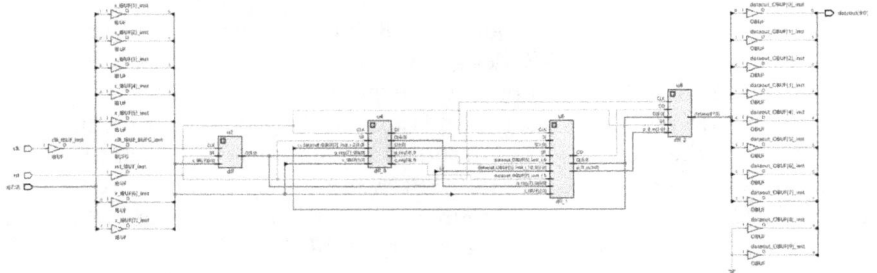

Figure 9.3 Synthesized RTL design of FIR filter.

From Table 9.1, it is observed that in the implementation process, 19 IO and LUTs as well as 28 FFs and 1 BUFG are utilized. The RTL schematic for the FIR filter is shown in Figure 9.2, while the synthesized RTL design of the FIR is depicted in Figure 9.3.

9.3 THERMAL PROPERTIES

This section will provide a quick overview of the device's thermal characteristics. When matching the impedance utilizing the SSTL IO standard, the thermal characteristics are taken into consideration. The thermal characteristics taken into account here include:

a. TM: This is a feature of a device that enables very low power consumption.
b. ϑJA: When 1 W of power is dissipated in the IC, the amount of heat created or a rise in temperature is specified.
c. JT: The highest operating temperature of an FPGA device's integrated gates.

The thermal properties for the SSTL IO for Zynq-7000 are described in Table 9.2. From Table 9.2, it is observed that the JT gets on increasing as the input voltage of the transmission line impedance matching increases [13–15]. The TM decreases as the input voltage gets on increasing for SSTL IO.

Table 9.2 Thermal properties for the SSTL IO for Zynq-7000

IO std	JT (°C)	TM (°C)	ϑJA (°C/W)
SSTL 135	82.5	17.5	11.5
SSTL 15	83.8	16.2	11.5
SSTL_18_I	125.0	−26.6	11.5
SSTL_18_II	125.0	−41.9	11.5

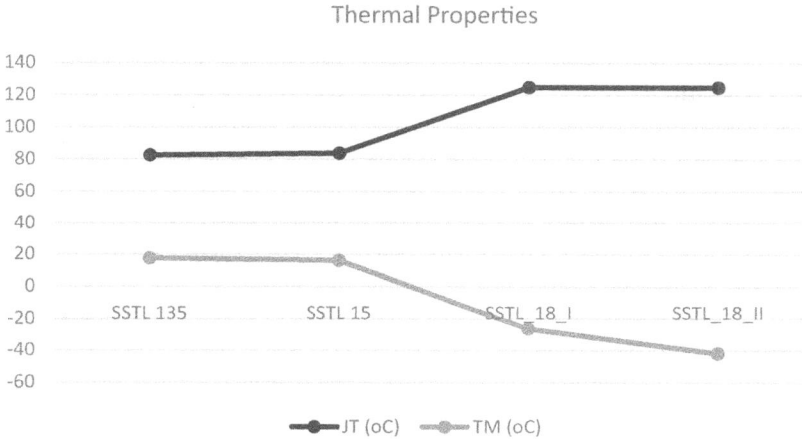

Figure 9.4 Thermal properties for the SSTL IO for Zynq-7000.

For the SSTL_18_I and SSTL_18_II IO, the TM value becomes negative. The ϑJA value remains constant to 11.5 for all SSTL IO. The variation in thermal properties is described in Figure 9.4.

9.4 POWER ANALYSIS

The biggest danger facing the world now comes from the energy usage of electronic devices and information technologies. The world is currently experiencing an energy shortage. As a result, the entire ecosystem works together to create a low-power solution for the information technology and electronics industries. The lifespan of the device is extended through lower power usage. In this section, we use the SSTL IO standard to determine the power usage of a FIR implementation on a Zynq-7000 device. The sum of the device's dynamic and static power consumption is the device's overall power consumption. Equation 9.1 provides a mathematical explanation of this [16–18]:

$$TP = SP + DP \tag{9.1}$$

where

TP = Total Power
SP = Static Power
DP = Dynamic Power

9.4.1 Power Analysis for SSTL 135

When the impedance is matched with the SSTL 135 IO, the total power consumption (TPC) observed for the device is 4.981. The DPC is 94% of the TPC which is 4.672 W, and SPC is 6% of the TPC which is 0.309 W. The on-chips power consumption for the device when impedance is matched with SSTL 135 IO is described in Figure 9.5.

From Figure 9.5, it can be clearly observed that the DPC is the combination of signal (S/G), logic (L/G), and IO power of the device. These are 0.262, 0.159, and 4.252 W, respectively. The graphical representation of the TPC for SSTL 135 is described in Figure 9.6.

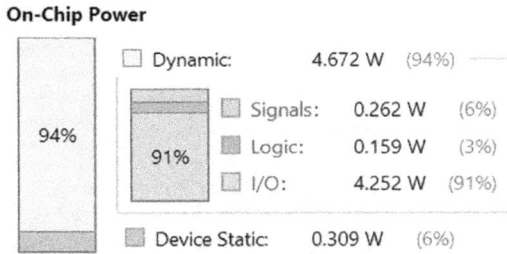

Figure 9.5 On-chips power consumption for SSTL 135 IO.

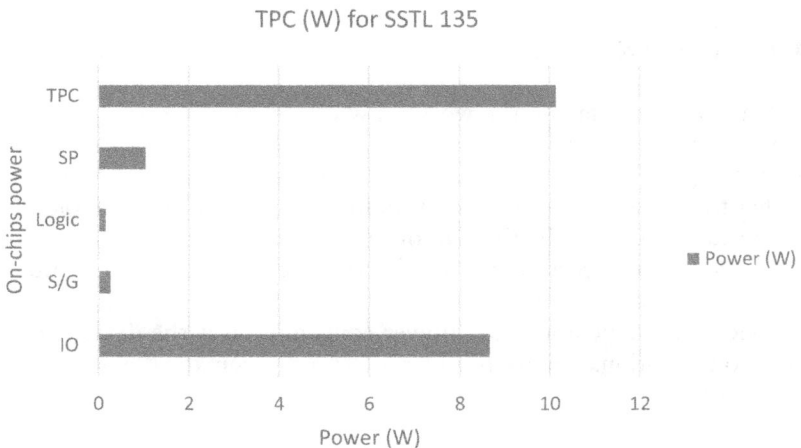

Figure 9.6 Graphical representation of the TPC for SSTL 135.

9.4.2 Power Analysis for SSTL 15

When the SSTL 15 IO is used for matching the impedance of the transmission line, it is observed that the DPC is 4.781 W, which is 94% of the TPC of the device. The SPC is just 6% of the TPC, which is 0.322 W. The TPC is the summation of SPC and DPC which is 5.103 W. The on-chips power consumption is described in Figure 9.7.

The DPC is the combination of S/G, L/G, and IO power of the device which are 0.262, 0.159, 4.360 W, respectively. The graphical representation of the TPC for SSTL 15 is presented in Figure 9.8.

9.4.3 Power Analysis for SSTL_18_1

When the impedance is matched with the SSTL_18_I IO, the total power consumption (TPC) observed for the device is 8.812 W. The DPC is 88% of the TPC which is 7.774 W, and SPC is 12% of the TPC which is 1.038 W.

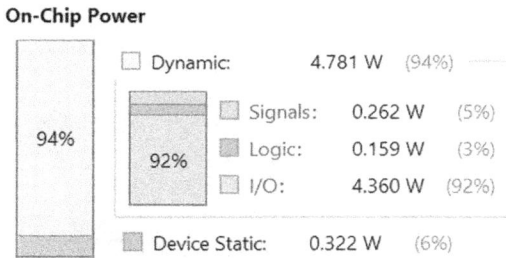

Figure 9.7 On-chips power consumption for SSTL 15.

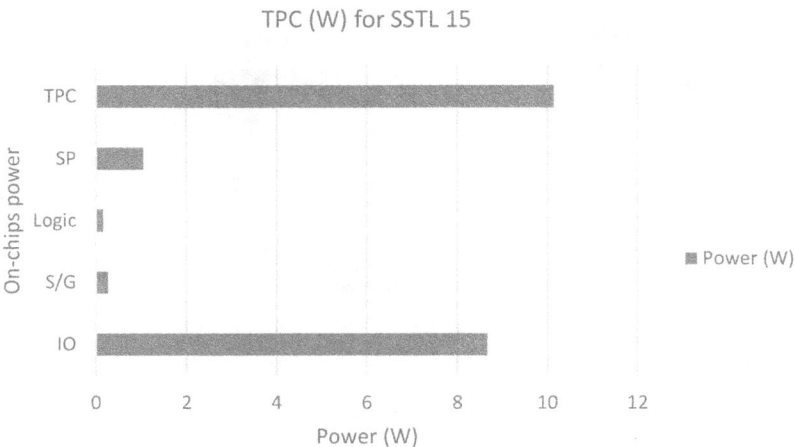

Figure 9.8 Graphical representation of the TPC for SSTL 15.

The on-chips power consumption for the device when impedance is matched with SSTL_18_I IO is described in Figure 9.9.

From Figure 9.10, it can be clearly observed that the DPC is the combination of signal (S/G), logic (L/G), and IO power of the device. These are 0.262, 0.159, and 7.354 W, respectively. The graphical representation of the TPC for SSTL_18_I is described in Figure 9.10.

9.4.4 Power Analysis for SSTL_18_II

When the SSTL_18_II IO is used for matching the impedance of the transmission line, it is observed that the DPC is 9.095 W, which is 90% of the TPC of the device. The SPC is 10% of the TPC, which is 1.042 W. The TPC is the summation of SPC and DPC which is 10.137 W. The on-chips power consumption is described in Figure 9.11.

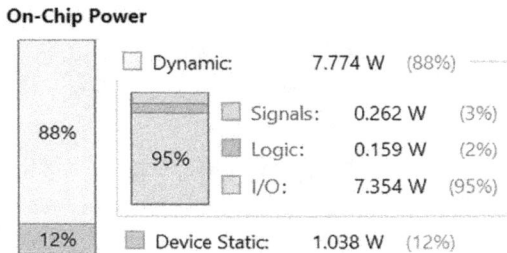

On-Chip Power

	Dynamic:	7.774 W	(88%)	
88%	Signals:	0.262 W	(3%)	
	95%	Logic:	0.159 W	(2%)
	I/O:	7.354 W	(95%)	
12%	Device Static:	1.038 W	(12%)	

Figure 9.9 On-chips power consumption for SSTL_18_I IO.

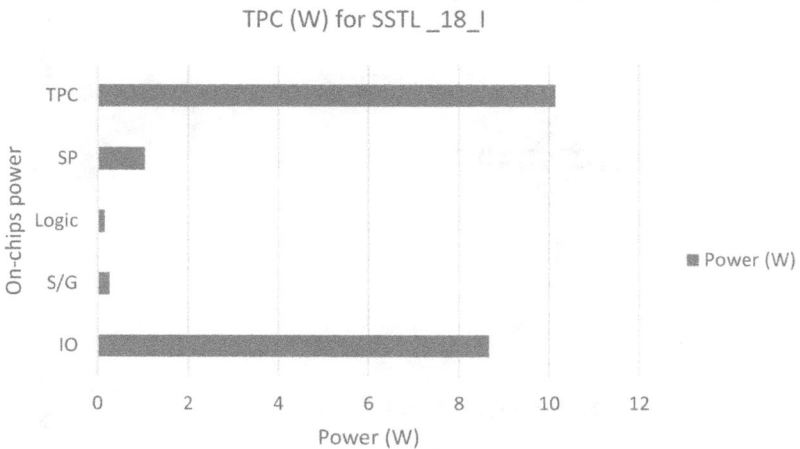

TPC (W) for SSTL _18_I

Figure 9.10 Graphical representation of the TPC for SSTL_18_I.

On-Chip Power

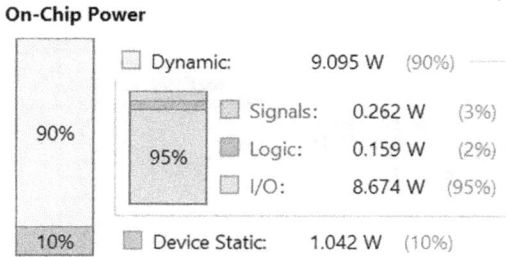

Figure 9.11 On-chips power consumption for SSTL_18_II.

TPC (W) for SSTL _18_II

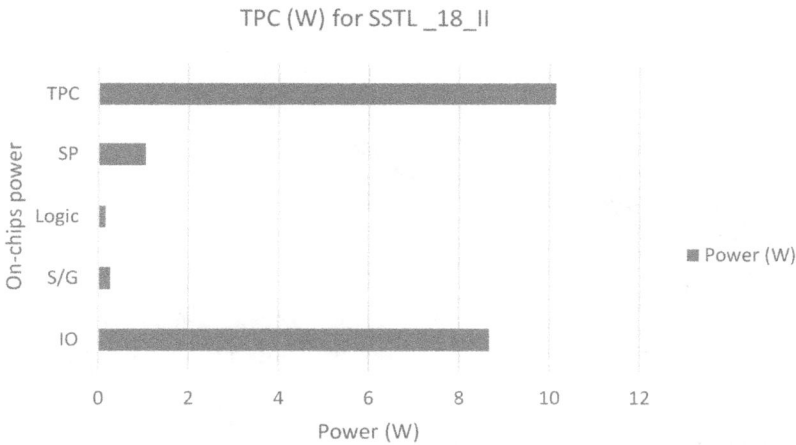

Figure 9.12 Graphical representation of the TPC for SSTL_18_II.

The DPC is the combination of S/G, L/G, and IO power of the device which are 0.2, 0.159, 8.674 W, respectively. The graphical representation of the TPC for SSTL_18_II is presented in Figure 9.12.

9.5 OBSERVATION AND ANALYSIS

This section will give insights about the analysis and observation from Section 9.4. From Section 9.4, it is observed that the TPC increases as the input voltage of the transmission line increases; i.e., TPC increases for SSTL IO is shown in Figure 9.13.

From Section 9.4, it is observed that TPC is minimum for SSTL 135 IO, and the TPC is maximum for SSTL_18_II IO. The TPC of the device for SSTL family IO is described in Figure 9.14.

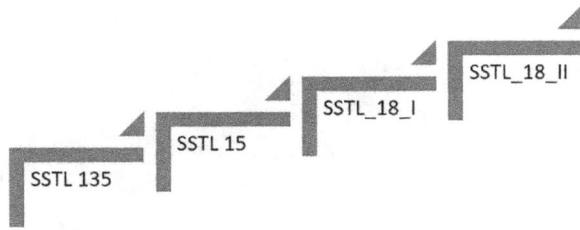

Figure 9.13 TPC increasing for SSTL IO.

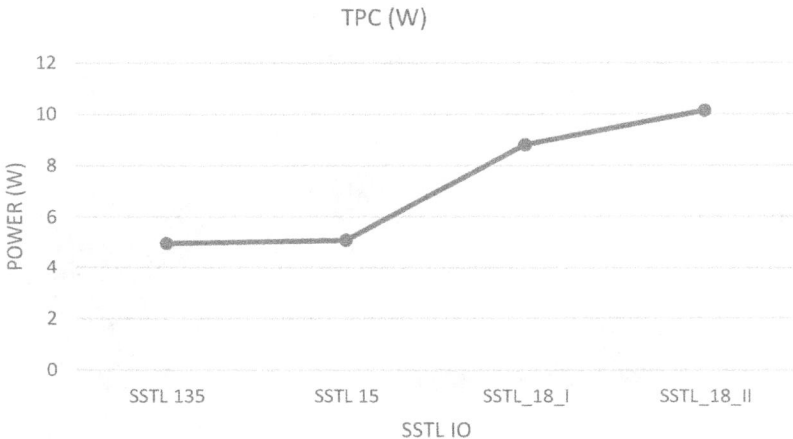

Figure 9.14 TPC of the device for SSTL family.

9.6 CONCLUSION

This chapter gives a detailed view of the TPC of the FIR filter on the ZYNQ-7000 FPGA. The FIR filter is one of the key components used in the world of signal processing, and apart from this, it has also several DSP applications. The FIR filter has a lot of several advantages over IIR filter. In this chapter, the power consumption of the FIR is analyzed over ZYNQ-7000. The design of FIR is implemented on VIVADO ISE. The power consumption is tried to be optimized with SSTL IO. IO standards are used to match the impedance of the input line with output line impedance. The power consumption of any device is optimized when there is a perfect impedance matching. Here various SSTL IOs are used to match the impedance. From Section 9.4, it is observed that as the input voltage of the IO standard increases, the TPC also increases. Therefore, the device gives the optimal power consumption when the input voltage is low. The FIR design on ZYNQ-7000 devices has the optimal power consumption when the impedance is matched with SSTL 135 IO. As far as future scope is concerned, the FIR design can be implemented over several other FPGAs of distinguished family such as Artix-7, Kinten-7, and Virtex-7. Also, the designs can be converted into ASIC design for better applications and uses.

REFERENCES

1. Koilpillai, R. David, and P. P. Vaidyanathan. "Cosine-modulated FIR filter banks satisfying perfect reconstruction." *IEEE Transactions on Signal Processing* 40, no. 4 (1992): 770–783.
2. Vaidyanathan, P., and Truong Nguyen. "Eigenfilters: A new approach to least-squares FIR filter design and applications including Nyquist filters." *IEEE Transactions on Circuits and Systems* 34, no. 1 (1987): 11–23.
3. Lim, Yong, and Sydney Parker. "FIR filter design over a discrete powers-of-two coefficient space." *IEEE Transactions on Acoustics, Speech, and Signal Processing* 31, no. 3 (1983): 583–591.
4. Hsiao, Shen-Fu, Jun-Hong Zhang Jian, and Ming-Chih Chen. "Low-cost FIR filter designs based on faithfully rounded truncated multiple constant multiplication/accumulation." *IEEE Transactions on Circuits and Systems II: Express Briefs* 60, no. 5 (2013): 287–291.
5. Davidson, Timothy N., Zhi-Quan Luo, and Jos F. Sturm. "Linear matrix inequality formulation of spectral mask constraints with applications to FIR filter design." *IEEE Transactions on Signal Processing* 50, no. 11 (2002): 2702–2715.
6. Hsiao, Shen-Fu, Jun-Hong Zhang Jian, and Ming-Chih Chen. "Low-cost FIR filter designs based on faithfully rounded truncated multiple constant multiplication/accumulation." *IEEE Transactions on Circuits and Systems II: Express Briefs* 60, no. 5 (2013): 287–291.
7. Vetterli, Martin, and Didier Le Gall. "Perfect reconstruction FIR filter banks: Some properties and factorizations." *IEEE Transactions on Acoustics, Speech, and Signal Processing* 37, no. 7 (1989): 1057–1071.
8. Mahapatra, Rajarshi, Yogesh Nijsure, Georges Kaddoum, Naveed Ul Hassan, and Chau Yuen. "Energy efficiency tradeoff mechanism towards wireless green communication: A survey." *IEEE Communications Surveys & Tutorials* 18, no. 1 (2015): 686–705.
9. Vereecken, Willem, Ward Van Heddeghem, Didier Colle, Mario Pickavet, and Piet Demeester. "Overall ICT footprint and green communication technologies." In *2010 4th International Symposium on Communications, Control and Signal Processing (ISCCSP)*, Limassol, Cyprus, pp. 1–6. IEEE, 2010.
10. Kumar, Tanesh, Bishwajeet Pandey, Teerath Das, and B. S. Chowdhry. "Mobile DDR IO standard based high performance energy efficient portable ALU design on FPGA." *Wireless Personal Communications* 76, no. 3 (2014): 569–578.
11. Amrbar, Mehran, Farokh Irom, Steven M. Guertin, and Greg Allen. "Heavy ion single event effects measurements of Xilinx Zynq-7000 FPGA." In *2015 IEEE Radiation Effects Data Workshop (REDW)*, Boston, MA, pp. 1–4. IEEE, 2015.

12. Al Kadi, Muhammed, Patrick Rudolph, Diana Gohringer, and Michael Hubner. "Dynamic and partial reconfiguration of Zynq 7000 under Linux." In *2013 International Conference on Reconfigurable Computing and FPGAs (ReConFig)*, Cancun, Mexico, pp. 1–5. IEEE, 2013.

13. Sharma, Rashmi, Bishwajeet Pandey, Vikas Jha, Siddharth Saurabh, and Sweety Dabas. "Input–output standard-based energy efficient UART design on 90ánm FPGA." In Sunil Kumar Muttoo (Ed.), *System and architecture*, pp. 139–150. Springer, Singapore, 2018.

14. Gupta, Isha, Swati Singh Garima, Harpreet Kaur, Deepshikha Bhatt, and Aamir Vohra. "28nm FPGA based power optimized UART Design using HSTL I/O standards." *Indian Journal of Science and Technology* 8, no. 17 (2015): 1–6.

15. Kumar, Keshav, Amanpreet Kaur, S. N. Panda, and Bishwajeet Pandey. "Effect of different nano meter technology-based FPGA on energy efficient UART design." In *2018 8th International Conference on Communication Systems and Network Technologies (CSNT)*, India, pp. 1–4. IEEE, 2018.

16. Kumar, Keshav, Amanpreet Kaur, Bishwajeet Pandey, and S. N. Panda. "Low power UART design using different nanometer technology-based FPGA." In 2018 *8th International Conference on Communication Systems and Network Technologies (CSNT)*, India, pp. 1–3. IEEE, 2018.

17. Pandey, Bishwajeet, and Ravikant Kumar. "Low voltage DCI based low power VLSI circuit implementation on FPGA." In *2013 IEEE Conference on Information & Communication Technologies*, India, pp. 128–131. IEEE, 2013.

18. Kaur, Harkinder, Harsh Sohal, and Jaiteg Singh. "Design and performance analysis of uart using Altera Quartus-II and Xilinx ISE 14.2." In *6th International Conference on Communication and Network Technologies*, India. 2016.

Chapter 10

HSTL-based FIR filter for GCC

LIST OF ABBREVIATIONS

AC	Alternating Current
ASIC	Application-Specific Integrated Circuit
BUFG	Global Buffers
Clk	Clocks
CMOS	Complementary Metal-Oxide-Semiconductor
DDR	Double Data Rate
DRAM	Dynamic Random-Access Memory
DP	Dynamic Power
FF	Flip-Flops
FPGA	Field Programmable Gate Array
GPS	Global Positioning System
GC	Green Computing
GCC	Green Communication Computing
G Comm.	Green Communication
HSTL	High-Speed Transceiver Logic
IC	Integrated Circuit
IO	Input Output
JT	Junction Temperature
LP	Leakage Power
LTE	Long-Term Evolution
LUT	Look Up Tables
RTL	Register Transfer Logic
S/G	Signal
SP	Static Power
TM	Thermal Margin
TP	Total Power
TPC	Total Power Consumption
UART	Universal Asynchronous Receiver Transmitter
WCDMA	Wideband Code Division Multiple Access
WiMAX	Worldwide Interoperability for Microwave Access
ϑJA	Effective Thermal Resistance to Air

DOI: 10.1201/9781003302872-10

10.1 INTRODUCTION

Increasing speed and reducing power consumption are crucial goals in many contemporary applications that use integrated circuits. These standards are met using cutting-edge complementary metal-oxide semiconductor (CMOS) technology thanks to the shrinking device sizes and low-voltage supply. The lower-voltage headroom and narrower dynamic range brought on by smaller device sizes and low-power supplies are challenges that many prior analog circuit architectures, such as analog-to-digital converters, struggle to overcome [1–3]. The time-domain approach, which is still relatively new, uses pulse width or time delay in place of voltage or current, as is the case with conventional processing systems, to process time. Time is hence a crucial component of time-domain circuits and systems. Time-domain design is a very promising design approach even with the limited technology available since it provides a broader dynamic range to power consumption trade-off. Time-domain systems have the advantage of using high-speed MOS transistors, which results in a shorter time delay and more precise time management [4,5]. Due to their composition of primarily CMOS digital building blocks for high-speed performance, time-domain designs have the advantages of improved dynamic range and time resolution when compared to analog voltage or current mode circuits in the same low-supply environment as well as improved power efficiency (gates, etc.); $z-1$ operators, signal adders, and signal multipliers are required for the production of filter coefficients in FIR implementations. Discrete-time/discrete-signal processing systems with a strong digital design foundation are the foundation of traditional FIR implementations. Flip-flops are used as delay components in the circuits that implement the basic operators, and digital logic gates are used to build adders and multipliers using digital logic. The FIR filter provides a ton of benefits in the communication and electronics fields [6,7]. The FIR filter can now be employed in additional fields thanks to technological advancements. Numerous applications consume a significant amount of power, which makes the circuitry large because the battery size may need to be increased. In addition, the lifespan may be shortened. As a result, the FIR implementation should be done in a way that uses less power. The low-power FIR filter implementation is the main topic of this chapter. The FIR design is built on an FPGA chip to make the design power-efficient. Electronics and communication technologies have several uses for the FPGA devices. Additionally, the power consumption of the built circuit can be optimized with the aid of FPGA devices [8–10]. The HSTL IO standard and Kintex-7 FPGA device are utilized to create a power-efficient FIR design.

The Kintex-7 FPGA is a 20Gb/s platform for high-bandwidth and high-performance applications that comes complete with all the hardware, resources, and intellectual property (IP) you'll need to complete your connection system development and assessment swiftly. The gate size of

Kintex-7 is 28 nm, and it optimizes the power of any implemented circuit in an efficient way. Designers can incorporate greater bandwidth and 12-bit digitally programmable analog with Kintex-7 FPGAs while still adhering to price and power constraints. The adaptable Kintex-7 devices are a great choice for uses like portable ultrasound equipment and next-generation communications thanks to the special 144GMACS digital signal processor (DSP) power. The 800Gbps full-duplex peak serial bandwidth of Kintex-7 FPGAs is combined with 9.8Gbps CPRI/OBSAI IP cores that are tailored for distributed baseband designs. The reconfigurable Kintex-7 devices can be quickly set up to handle a variety of air interfaces, including LTE, WiMAX, and WCDMA. Applications including flat panel displays, video over IP, and 3G, and 4G wireless are all excellent candidates for the Kintex-7 family [11–13].

10.2 FIR FILTER IMPLEMENTATION ON KINTEX-7 FPGA DEVICE

We used the Kintex-7 FPGA device to build a FIR filter model that is power-efficient. The VIVADO ISE is used to implement the FIR design on the same hardware. HSTL IOs are used to optimize power usage. The internal circuit's IO standard matches the transmission line impedance, aiming to reduce power consumption. Resources from the FPGA device, such as IO, FF, LUTs, and BUFG, are used to implement the design. Table 10.1 and Figure 10.1 demonstrate and describe how resources were used to construct the FIR [14,15].

From Table 10.1, it is observed that in the implementation process, 19 IO and LUTs as well as 28 FFs and 1 BUFG are utilized. The RTL schematic for the FIR filter is shown in Figure 10.2, while the synthesized RTL design of the FIR is depicted in Figure 10.3.

10.3 THERMAL PROPERTIES

This section will provide a quick overview of the device's thermal characteristics. When matching the impedance utilizing the HSTL IO standard,

Table 10.1 Resource utilization for designing FIR design

Resources	Available	Utilization	Utilization%
LUT	19	10,1400	0.02
IO	19	400	4.750
BUFG	1	32	3.13
FF	28	20,2800	0.01

Resource Utilization

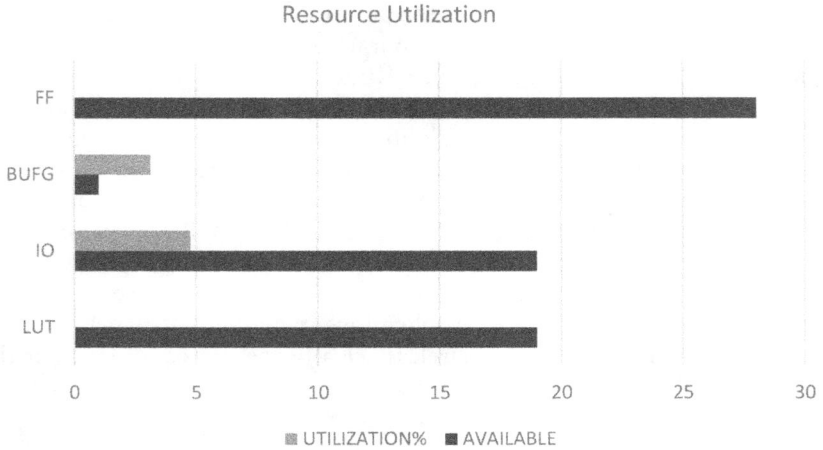

Figure 10.1 Resource utilization for designing FIR design.

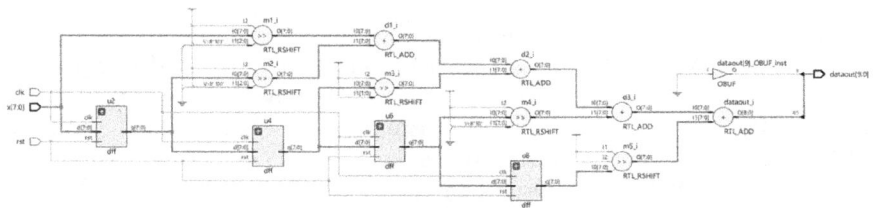

Figure 10.2 RTL schematic of FIR filter.

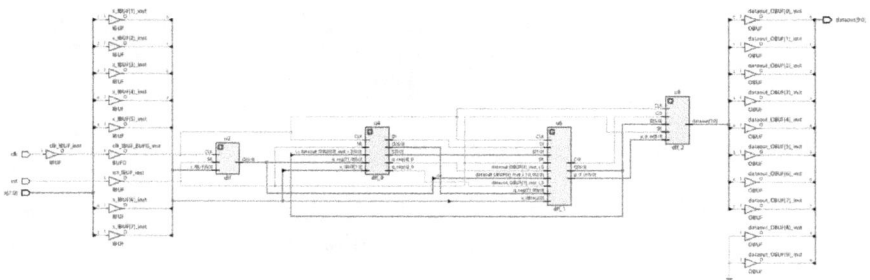

Figure 10.3 Synthesized RTL design of FIR filter.

the thermal characteristics are taken into consideration [16]. The thermal characteristics taken into account here include:

a. TM: This is a feature of a device that enables very low power consumption.

b. ϑJA: When 1 W of power is dissipated in the IC, the amount of heat created or a rise in temperature is specified.

c. JT: The highest operating temperature of an FPGA device's integrated gates.

The thermal properties for the HSTL IO for Kintex-7 are described in Table 10.2. The variation in thermal properties is described in Figure 10.4. From Table 10.2, it is observed that JT is at max. for HSTL_18_I IO, while it is optimal for HSTL_II. The ϑJA (°C/W) is a constant for all the IOs (1.9), while TM is optimal for HSTL_II IO.

10.4 POWER ANALYSIS

The biggest danger facing the world now comes from the energy usage of electronic devices and information technologies. The world is currently experiencing an energy shortage. As a result, the entire ecosystem works

Table 10.2 Thermal properties for the HSTL IO for Kintex-7

IO std	JT (°C)	TM (°C)	ϑJA (°C/W)
HSTL_I	38.5	61.5	1.9
HSTL_II	32.8	67.2	1.9
HSTL_18_I	41.4	58.6	1.9
HSTL_18_II	33.5	66.6	1.9
HSTL_I_12	35.8	64.2	1.9

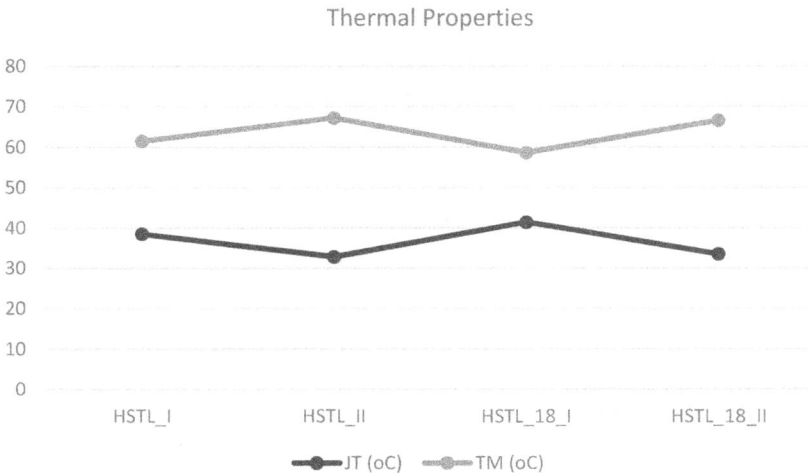

Figure 10.4 Thermal properties for the HSTL IO for Kintex-7.

together to create a low-power solution for the information technology and electronics industries. The lifespan of the device is extended through lower power usage. In this section, we use the HSTL IO standard to determine the power usage of a FIR implementation on a Kintex-7 device. The sum of the device's dynamic and static power consumption is the device's overall power consumption [17,18]. Equation 10.1 provides a mathematical explanation of this:

$$TP = SP + DP \tag{10.1}$$

where

$TP = $ Total Power
$SP = $ Static Power
$DP = $ Dynamic Power

10.4.1 Power analysis for HSTL_I

When the impedance is matched with the HSTL_I IO, the total power consumption (TPC) observed for the device is 6.993. The DPC is 98% of the TPC which is 6.846 W, and SPC is 2% of the TPC which is 0.146 W. The on-chips power consumption for the device when impedance is matched with HSTL_I IO is described in Figure 10.5.

From Figure 10.6, it can be clearly observed that the DPC is the combination of signal (S/G), logic (L/G), and IO power of the device. These are 0.258, 0.151, and 6.437 W, respectively. The graphical representation of the TPC for HSTL_I is described in Figure 10.6.

10.4.2 Power analysis for HSTL_II

When the HSTL_II IO is used for matching the impedance of the transmission line, it is observed that the DPC is 3.948 W, which is 97% of the TPC of the device. The SPC is just 3% of the TPC, which is 0.130 W. The TPC is the summation of SPC and DPC which is 4.078 W. The on-chips power consumption is described in Figure 10.7.

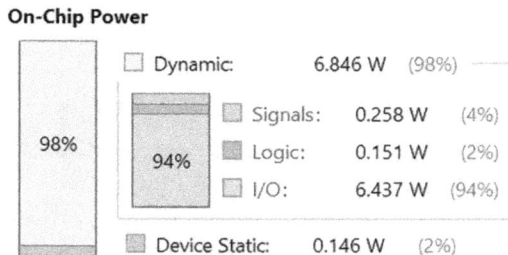

Figure 10.5 On-chips power consumption for HSTL_I IO.

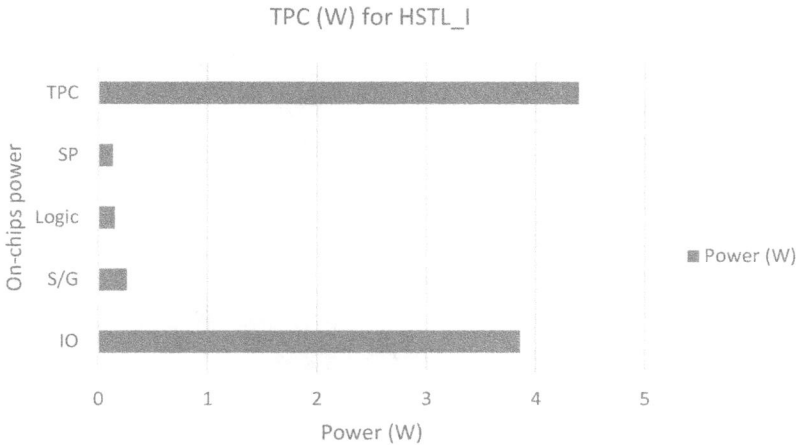

Figure 10.6 Graphical representation of the TPC for HSTL_I.

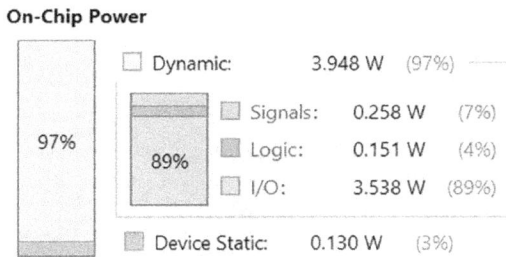

Figure 10.7 On-chips power consumption for HSTL_II.

The DPC is the combination of S/G, L/G, and IO power of the device which are 0.258, 0.151, 3.538 W, respectively. The graphical representation of the TPC for HSTL_II is presented in Figure 10.8.

10.4.3 Power analysis for HSTL_I_I2

When the impedance is matched with the HSTL_I_12 IO, the total power consumption (TPC) observed for the device is 5.612 W. The DPC is 98% of the TPC which is 5.473 W, and SPC is 2% of the TPC which is 0.139 W. The on-chips power consumption for the device when impedance is matched with HSTL_I_12 IO is described in Figure 10.9.

From 10.9, it can be clearly observed that the DPC is the combination of signal (S/G), logic (L/G), and IO power of the device. These are 0.258, 0.151, and 5.046 W, respectively. The graphical representation of the TPC for HSTL_I_12 is described in Figure 10.10.

TPC (W) for HSTL_II

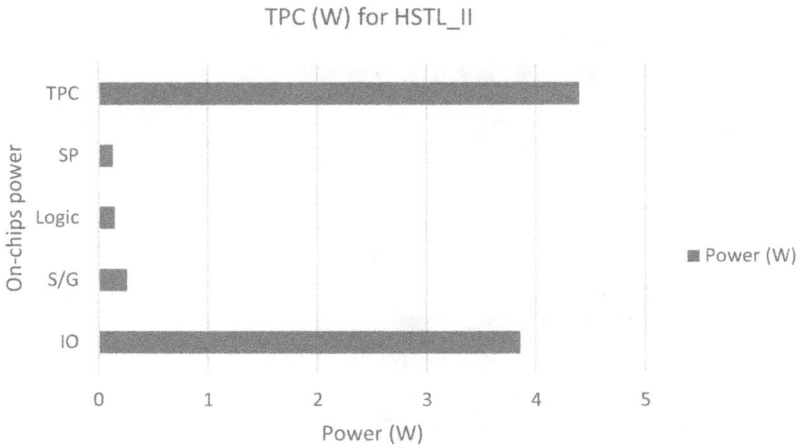

Figure 10.8 Graphical representation of the TPC for HSTL_II.

On-Chip Power

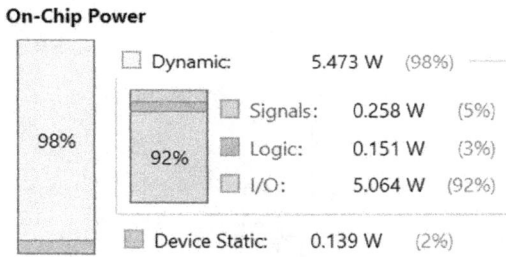

Dynamic:	5.473 W	(98%)
Signals:	0.258 W	(5%)
Logic:	0.151 W	(3%)
I/O:	5.064 W	(92%)
Device Static:	0.139 W	(2%)

Figure 10.9 On-chips power consumption for HSTL_I_I2 IO.

TPC (W) for HSTL_I_12

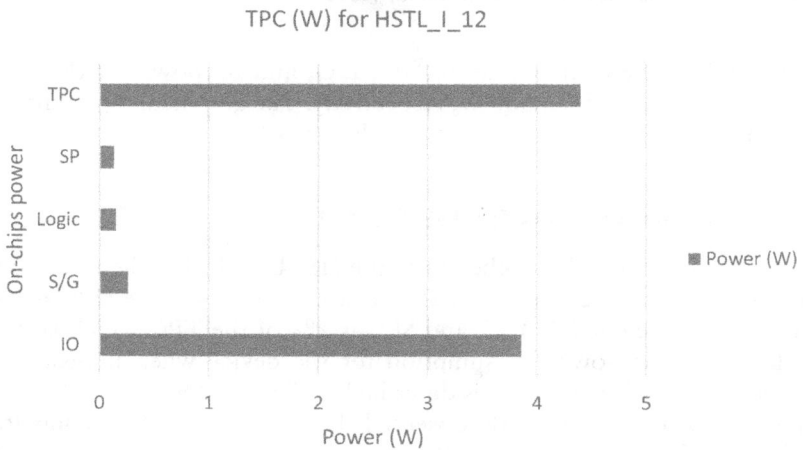

Figure 10.10 Graphical representation of the TPC for HSTL_I_I2.

10.4.4 Power analysis for HSTL_I_18

When the HSTL_I_18 IO is used for matching the impedance of the transmission line, it is observed that the DPC is 8.355 W, which is 98% of the TPC of the device. The SPC is 2% of the TPC, which is 0.157 W. The TPC is the summation of SPC and DPC which is 8.511 W. The on-chips power consumption is described in Figure 10.11.

The graphical representation of the TPC for HSTL_18_II is presented in Figure 10.12.

10.4.5 Power analysis for HSTL_II_18

When the HSTL_II_18 IO is used for matching the impedance of the transmission line, it is observed that the DPC is 4.265 W, which is 97% of the TPC of the device. The SPC is 3% of the TPC, which is 0.131 W. The TPC is the summation of SPC and DPC which is 4.397 W. The on-chips power consumption is described in Figure 10.13.

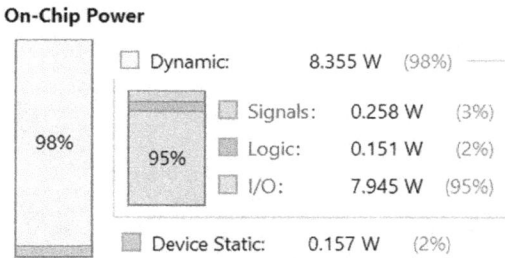

On-Chip Power

☐ Dynamic:	8.355 W	(98%)	
▣ Signals:	0.258 W	(3%)	
▣ Logic:	0.151 W	(2%)	
☐ I/O:	7.945 W	(95%)	
▣ Device Static:	0.157 W	(2%)	

Figure 10.11 On-chips power consumption for HSTL_I_18.

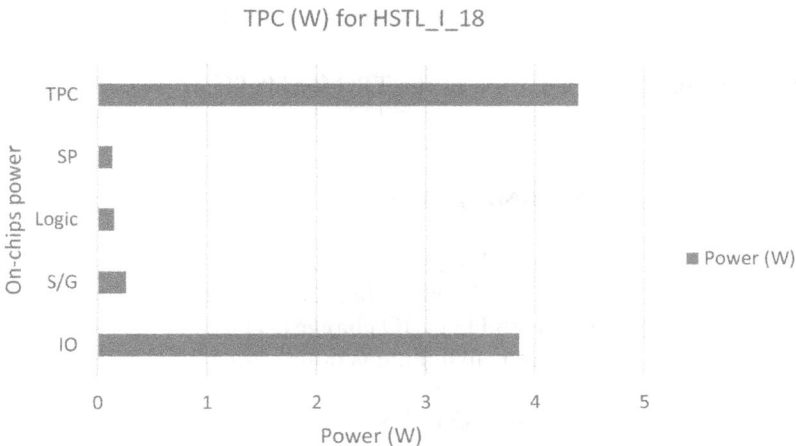

TPC (W) for HSTL_I_18

Figure 10.12 Graphical representation of the TPC for HSTL_I_18.

On-Chip Power

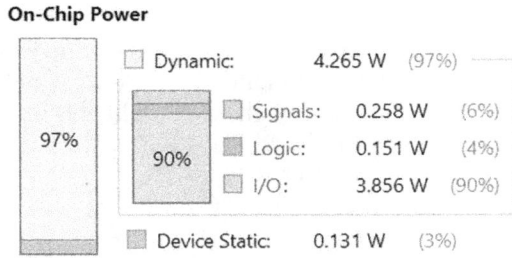

Dynamic:	4.265 W	(97%)
Signals:	0.258 W	(6%)
Logic:	0.151 W	(4%)
I/O:	3.856 W	(90%)
Device Static:	0.131 W	(3%)

Figure 10.13 On-chips power consumption for HSTL_II_18.

TPC (W) for HSTL_II_18

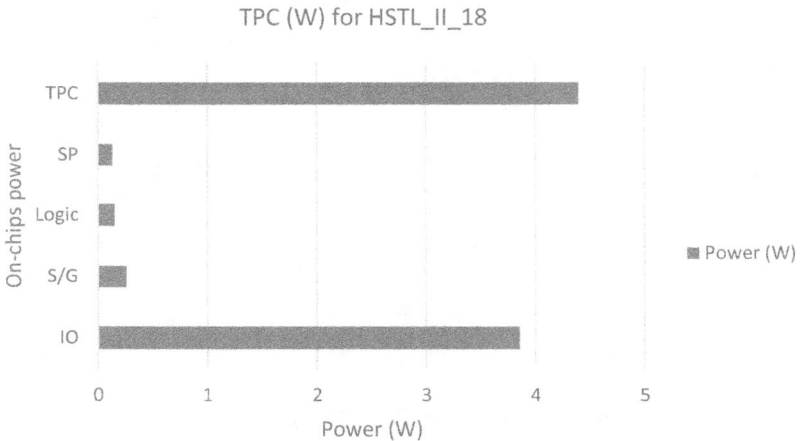

Figure 10.14 Graphical representation of the TPC for HSTL_II_18.

The graphical representation of the TPC for HSTL_II_18 is presented in Figure 10.14.

10.5 OBSERVATION AND ANALYSIS

This section will give insights about the analysis and observation from Section 10.4. From Section 10.4, it is observed that the TPC of the devices varies as the impedance matching with HSTL IO changes. The device consumes more power when the impedance is matching is done with HSTL_I_18 IO, while the device performs under low-power condition with HSTL_II IO. The TPC of the device for HSTL family IO is described in Figure 10.15.

TPC (W)

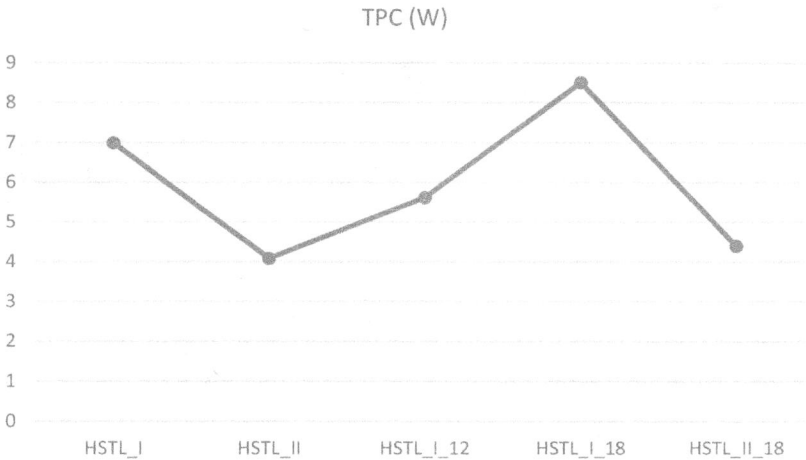

Figure 10.15 TPC of the device for HSTL family.

10.6 CONCLUSION

This chapter examines the TPC of the FIR Filter on the Kintex-7 FPGA in detail. The FIR filter is an important component in the field of signal processing, and it also has various DSP applications. The FIR filter has numerous advantages over the IIR filter. The power consumption of the FIR over Kintex-7 is examined in this chapter. The FIR design is implemented on VIVADO ISE. With HSTL IO, the power consumption is attempted to be optimized. IO standards are used to match the input line impedance to the output line impedance. When there is perfect impedance matching, the power consumption of any device is optimized. To match the impedance, several HSTL IOs are used. From Section 10.4, it is observed that the device consumes optimal power when the impedance matching for the FIR filter is matched with HSTL_II IO. The devices give maximum power dissipation when the impedance is matched with HSTL_I_18 IO. There is a change of 108.705% in TPC for HSTL_II (Optimal) to HSTL_I_18 (Maximum) IO. In terms of future potential, the FIR architecture can be implemented on a variety of different FPGAs from the renowned family, including the Artix-7, Zynq-7000, and Virtex-7. The designs can also be transformed into ASIC designs for improved applications and usage.

REFERENCES

1. Koilpillai, R. David, and P. P. Vaidyanathan. "Cosine-modulated FIR filter banks satisfying perfect reconstruction." *IEEE Transactions on Signal Processing* 40, no. 4 (1992): 770–783.
2. Vaidyanathan, P., and Truong Nguyen. "Eigenfilters: A new approach to least-squares FIR filter design and applications including Nyquist filters." *IEEE Transactions on Circuits and Systems* 34, no. 1 (1987): 11–23.
3. Lim, Yong, and Sydney Parker. "FIR filter design over a discrete powers-of-two coefficient space." *IEEE Transactions on Acoustics, Speech, and Signal Processing* 31, no. 3 (1983): 583–591.
4. Hsiao, Shen-Fu, Jun-Hong Zhang Jian, and Ming-Chih Chen. "Low-cost FIR filter designs based on faithfully rounded truncated multiple constant multiplication/accumulation." *IEEE Transactions on Circuits and Systems II: Express Briefs* 60, no. 5 (2013): 287–291.
5. Davidson, Timothy N., Zhi-Quan Luo, and Jos F. Sturm. "Linear matrix inequality formulation of spectral mask constraints with applications to FIR filter design." *IEEE Transactions on Signal Processing* 50, no. 11 (2002): 2702–2715.
6. Hsiao, Shen-Fu, Jun-Hong Zhang Jian, and Ming-Chih Chen. "Low-cost FIR filter designs based on faithfully rounded truncated multiple constant multiplication/accumulation." *IEEE Transactions on Circuits and Systems II: Express Briefs* 60, no. 5 (2013): 287–291.
7. Vetterli, Martin, and Didier Le Gall. "Perfect reconstruction FIR filter banks: Some properties and factorizations." *IEEE Transactions on Acoustics, Speech, and Signal Processing* 37, no. 7 (1989): 1057–1071.
8. Mahapatra, Rajarshi, Yogesh Nijsure, Georges Kaddoum, Naveed Ul Hassan, and Chau Yuen. "Energy efficiency tradeoff mechanism towards wireless green communication: A survey." *IEEE Communications Surveys & Tutorials* 18, no. 1 (2015): 686–705.
9. Vereecken, Willem, Ward Van Heddeghem, Didier Colle, Mario Pickavet, and Piet Demeester. "Overall ICT footprint and green communication technologies." In *2010 4th International Symposium on Communications, Control and Signal Processing (ISCCSP)*, Limassol, Cyprus, pp. 1–6. IEEE, 2010.
10. Kumar, Tanesh, Biswajeet Pandey, Teerath Das, and B. S. Chowdhry. "Mobile DDR IO standard based high performance energy efficient portable ALU design on FPGA." *Wireless Personal Communications* 76, no. 3 (2014): 569–578.
11. Kintex-7 FPGA family. https://www.xilinx.com/products/silicon-devices/fpga/kintex-7.html
12. Kuang, Jie, Yonggang Wang, Qiang Cao, and Chong Liu. "Implementation of a high precision multi-measurement time-to-digital convertor on a Kintex-7 FPGA." *Nuclear Instruments and Methods in Physics Research Section A: Accelerators, Spectrometers, Detectors and Associated Equipment* 891 (2018): 37–41.
13. Wirthlin, M. J., H. Takai, and A. Harding. "Soft error rate estimations of the Kintex-7 FPGA within the ATLAS Liquid Argon (LAr) Calorimeter." *Journal of Instrumentation* 9, no. 01 (2014): C01025.

14. Kaur, Ravinder, Jagdish Kumar, Sumita Nagah, Bishwajeet Pandey, and Kavita Goswami. "IO Standard based low power memory design and implementation on FPGA." In *2015 2nd International Conference on Computing for Sustainable Global Development (INDIACom)*, India, pp. 1501–1505. IEEE, 2015.

15. Sharma, Rashmi, Bishwajeet Pandey, Vikas Jha, Siddharth Saurabh, and Sweety Dabas. "Input–output standard-based energy efficient UART design on 90nm FPGA." In Sunil Kumar Muttoo (Ed.), *System and architecture*, pp. 139–150. Springer, Singapore, 2018.

16. Kumar, Keshav, Amanpreet Kaur, and K. R. Ramkumar. "Effective data transmission with UART on Kintex-7 FPGA." In *2020 12th International Conference on Computational Intelligence and Communication Networks (CICN)*, India, pp. 492–497. IEEE, 2020.

17. Aggarwal, Arushi, Bishwajeet Pandey, Sweety Dabbas, Achal Agarwal, and Siddharth Saurabh. "LVCMOS-based low-power thermal-aware energy-proficient vedic multiplier design on different FPGAs." In Sunil Kumar Muttoo (Ed.), *System and architecture*, pp. 115–122. Springer, Singapore, 2018.

18. Kumar, Abhishek, Bishwajeet Pandey, DM Akbar Hussain, Mohammad Atiqur Rahman, Vishal Jain, and Ayoub Bahanasse. "Frequency scaling and high-speed transceiver logic based low power UART design on 45nm FPGA." In *2019 11th International Conference on Computational Intelligence and Communication Networks (CICN)*, India, pp. 88–92. IEEE, 2019.

Chapter 11

MOBILE DDR-based FIR filter for GCC

LIST OF ABBREVIATIONS

AC	Alternating Current
ASIC	Application-Specific Integrated Circuit
BUFG	Global Buffers
Clk	Clocks
CMOS	Complementary Metal-Oxide-Semiconductor
DDR	Double Data Rate
DRAM	Dynamic Random-Access Memory
DP	Dynamic Power
FF	Flip-Flops
FPGA	Field Programmable Gate Array
GC	Green Computing
GCC	Green Communication Computing
G Comm.	Green Communication
GPS	Global Positioning System
IC	Integrated Circuit
IO	Input Output
JT	Junction Temperature
LP	Leakage Power
LUT	Look Up Tables
RTL	Register Transfer Logic
S/G	Signal
SP	Static Power
TM	Thermal Margin
TP	Total Power
TPC	Total Power Consumption
UART	Universal Asynchronous Receiver Transmitter
ϑJA	Effective thermal resistance to air

DOI: 10.1201/9781003302872-11

II.I INTRODUCTION

In the era where daily upgradation in the field of technologies is taking place across the globe, the planet is most concerned with the two major topics: the first is better speed, and the second is low power consumption of the devices. Day by day, the size of the communicating devices used for communicating is getting decreased, so such devices can't allow the big battery backup to increase the lifetime of the device. Therefore, such devices don't allow more dynamic range [1,2].

The time-domain technique is a fairly new way to process time that uses time delay, time difference, or pulse width instead of voltage or current, as was done in the past. So, time is the thing that matters in time-domain circuits and systems. Even for a technology that is so small, time-domain design is a very promising way to design because it gives a better balance between dynamic range and power consumption. Time-domain systems have an advantage because they use high-speed MOS devices [3,4]. This means they have a shorter time delay and can process time more accurately. The transfer of signals for the applications related to image processing, biomedical applications, interfaces, and wireless communications is generally done in time domain. These signals can be filtered in time domain using the FIR/IIR filters. The FIR filters are designed using some digital components such as adder, multiplexers, FF, basic gates, and some components of z^{-1} operators [4,5]. The realization of such system is regarded as DT-CSP system. The implementation of FIR using the digital circuits is quite reliable, but in doing so, we have to do a trade-off between the area (on-chips area), frequency, and power consumption [6].

This chapter is about designing a FIR filter using Kintex-7 FPGA SoC which can promote the ethics of GC. The proposed time-domain design offers various benefits that contribute to the robust implementation of FIR filters such as:

a. Less complex circuit
b. Excellent precision in time storage
c. Synchronization clock pulses
d. A modular design.

In addition to the abovementioned benefits, the FIR design is implemented on FPGA to make the design power-efficient so that the people across the globe can be aware of the concepts of GC [7,8]. To do so, the FIR design is implemented over Kintex-7 FPGA device, while the impedance matching of the circuitry has been done using the MOBILE DDR IO standard. The power analysis of the implemented design has been calculated for different frequencies ranging from 100 MHz to 5 GHz. The implementation has been done using the VIVADO ISE suite [9].

11.2 IMPLEMENTATION OF FIR DESIGN ON KINTEX-7 FPGA

The GC model of the FIR design is implemented on the Kintex-7 device using VIVADO ISE suite. The power calculation of the design has been tested for various frequencies ranging from 100 MHz to 5 GHz, while the impedance of the circuit is matched using MOBILE DDR IO. The impedance is matched in order to optimize the power consumption of the device. In order to implement the design on FPGA, many FPGA resources have been utilized such as FF, IO, BUFG, and LUTs [10,11]. The utilization of FPGA resources is demonstrated in Table 11.1 and graphically described in Figure 11.1.

From Table 11.1, it is observed that in the implementation process 19 IO and LUTs as well as 28 FFs and 1 BUFG are utilized. The RTL schematic for the FIR filter is shown in Figure 11.2, while the synthesized RTL design of the FIR is depicted in Figure 11.3.

Table 11.1 Resource utilization for designing FIR design

Resources	Available	Utilization	Utilization (%)
LUT	19	1,01,400	0.02
IO	19	400	4.750
BUFG	1	32	3.13
FF	28	2,02,800	0.01

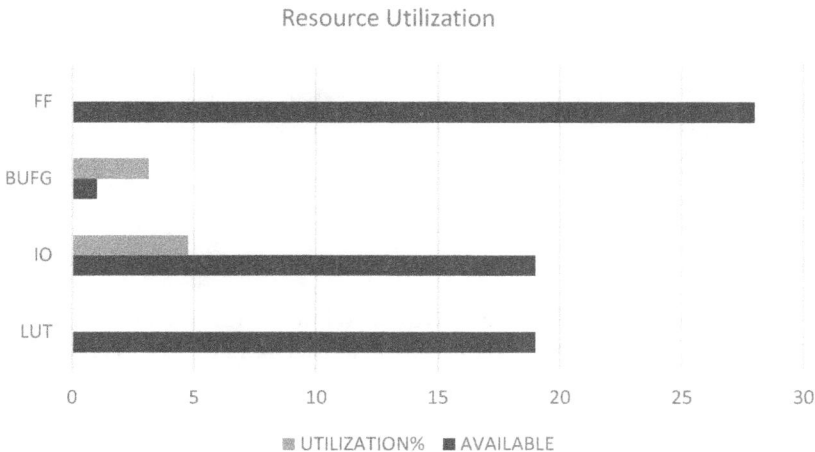

Figure 11.1 Resource utilization for designing FIR design.

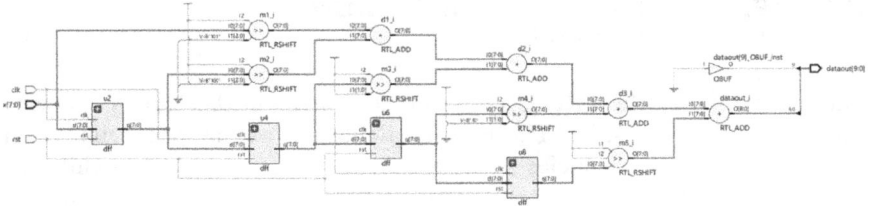

Figure 11.2 RTL schematic of FIR filter.

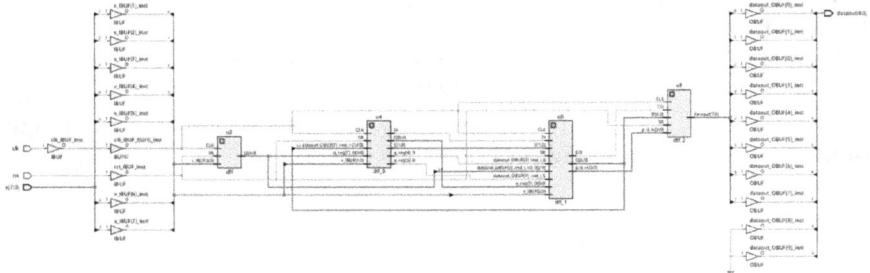

Figure 11.3 Synthesized RTL design of FIR filter.

11.3 THERMAL PROPERTIES

This section will provide an overview of the thermal characteristics of the device. The thermal properties are taken into account while matching impedance using the MOBILE DDR IO standard. The following thermal characteristics are taken into account:

a. TM: This is a feature of a device that enables very low power consumption.
b. ϑJA: When 1 W of power is dissipated in the IC, the amount of heat created or a rise in temperature is specified.
c. JT: The highest operating temperature of an FPGA device's integrated gates.

The thermal properties for the MOBILE DDR IO for various frequencies are described in Table 11.2. The variation in thermal properties is described in Figure 11.4. From Table 11.2, it is observed that for the frequency of 100 MHz to 1 GHz, the thermal properties like JT, TM, and ϑJA remain the same [12,13]. The slight change in JT and TM is only observed for 5 GHz frequency.

Table 11.2 Thermal properties for the MOBILE DDR
IO for Kintex-7

Frequency	JT (°C)	TM (°C)	ϑJA (°C/W)
100 MHz	25.3	74.7	1.9
300 MHz	25.3	74.7	1.9
500 MHz	25.3	74.7	1.9
1 GHz	25.3	74.7	1.9
5 GHz	25.5	74.5	1.9

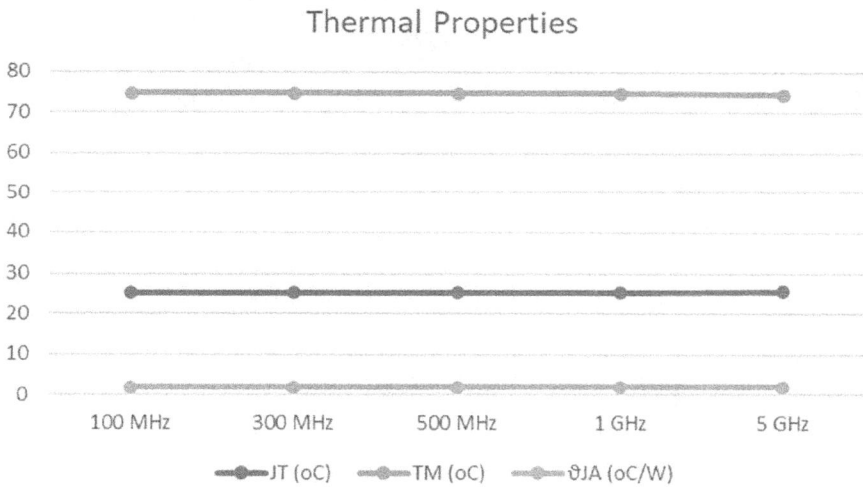

Figure 11.4 Thermal properties for the MOBILE DDR IO for Kintex-7.

11.4 POWER ANALYSIS

In the current state of the globe, the power consumption of gadgets and information technology poses the greatest threat. The entire planet is experiencing an energy deficit. Thus, the entire ecosystem collaborates to produce a low-power solution for the electronics and information technology industries. Reduced power consumption prolongs the device's life. In this part, we calculate the power consumption of a FIR filter implementation on a Kintex-7 device using the MOBILE DDR IO standard. The gadget's total power consumption is the sum of its dynamic and static power consumption [14,15]. This is expressed mathematically in Equation 11.1:

$$TP = SP + DP \tag{11.1}$$

where

TP = Total Power
SP = Static Power
DP = Dynamic Power

The device SP is the total of the device's Clocks (clk), IO, Logic, and Signa (S/G) power, while the DP is the device's leakage power (LP).

11.4.1 Power analysis for 100 MHz frequency

When the frequency is tuned to 100 MHz value, the TPC of the device calculated is 0.151 W, which is the summation of DP (0.039 W; 26% of TPC) and SP (0.112 W; 74% of TPC). The TPC for 100 MHz is illustrated in Figure 11.5.

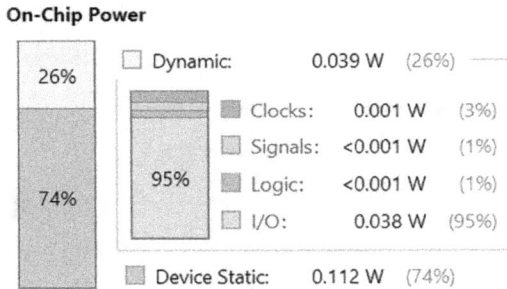

On-Chip Power

	Dynamic:	0.039 W	(26%)
26%			
	Clocks:	0.001 W	(3%)
	Signals:	<0.001 W	(1%)
95% Logic:	<0.001 W	(1%)
74%	I/O:	0.038 W	(95%)
	Device Static:	0.112 W	(74%)

Figure 11.5 TPC for MOBILE DDR IO at 100 MHz.

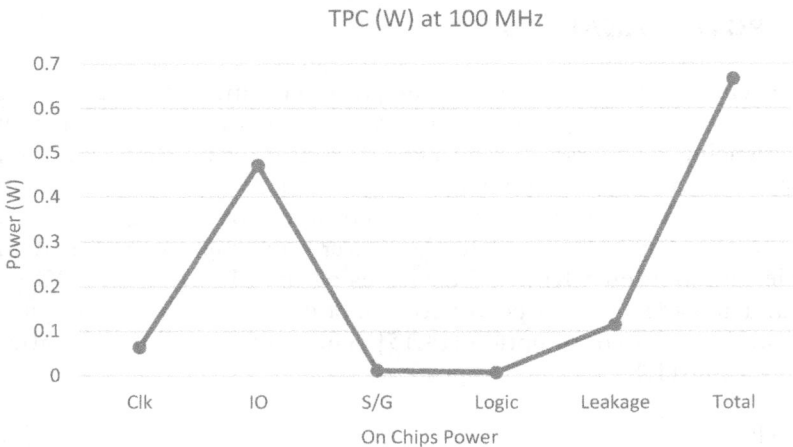

Figure 11.6 Representation of on-chips power for 100 MHz.

The DP is the total of clk, IO, logic, and S/G power, which are, respectively, 0.001, 0.038, <0.001, and <0.001 W. The SP is 0.112 W. The on-chips power for 100 MHz frequency is shown in Figure 11.6.

11.4.2 Power analysis for 300 MHz frequency

When the frequency is tuned to 300 MHz for matching the impedance of the device the TP consumption of the device observed is 0.172 W. The device SP is 0.112 W which is 65% of TP, and DP is 0.060 W which is 35% of TP consumption. The power consumption of the FIR filter for 300 MHz is illustrated in Figure 11.7. The graphical representation of TPC is described in Figure 11.8.

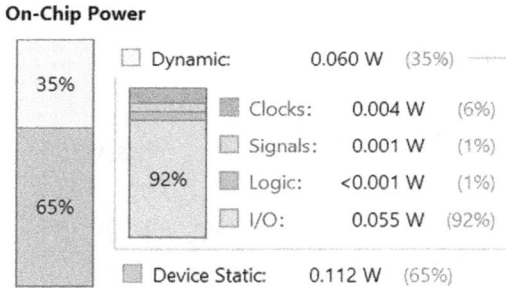

Figure 11.7 TPC for MOBILE DDR IO at 300 MHz.

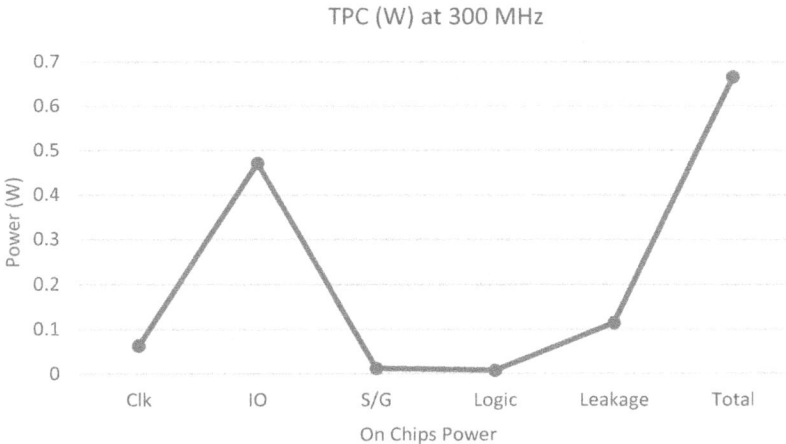

Figure 11.8 Representation of on-chips power for 300 MHz.

11.4.3 Power analysis for 500 MHz frequency

As the frequency of operation is increased to the value of 500 MHz, the TP of the device measured is 0.193 W, which is the cumulative sum of SP (0.112 W); 58% of the TP and DP (0.081 W); and 42% of the TP. The DP is the total of clk, S/G, logic, and IO power, which are 0.006, 0.001, 0.001, and 0.073 W, respectively. The representation of the total on-chips power is shown in Figure 11.9, and the graphical representation is described in Figure 11.10.

11.4.4 Power analysis for 1 GHz frequency

When the frequency is tuned to 1 GHz for the impedance matching, it is observed that the TPC is 0.246 W, which is the summation of DP (0.134 W) and SP is (0.112 W). The TP consumption for 1 GHz is illustrated in Figure 11.11.

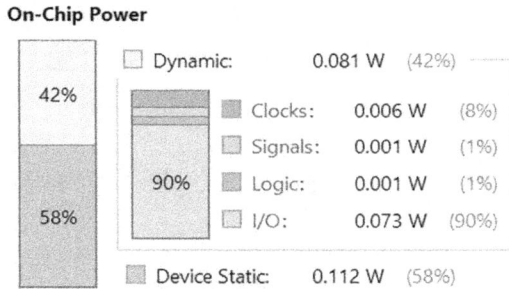

Figure 11.9 TPC of MOBILE DDR IO at 500 MHz frequency.

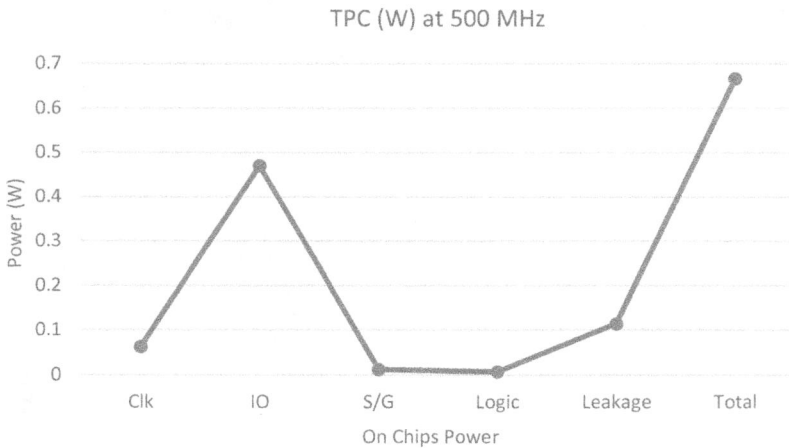

Figure 11.10 Representation of on-chips power for 500 MHz.

On-Chip Power

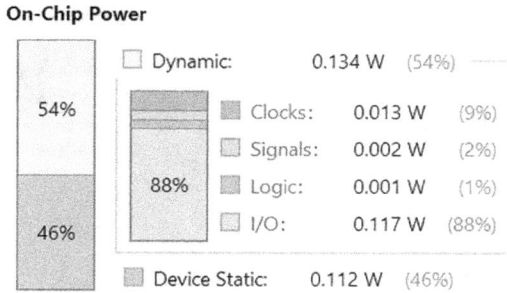

☐ Dynamic:	0.134 W	(54%)
☐ Clocks:	0.013 W	(9%)
☐ Signals:	0.002 W	(2%)
☐ Logic:	0.001 W	(1%)
☐ I/O:	0.117 W	(88%)
☐ Device Static:	0.112 W	(46%)

Figure 11.11 TP consumption for MOBILE DDR IO at 1 GHz frequency.

TPC (W) at 1 GHz

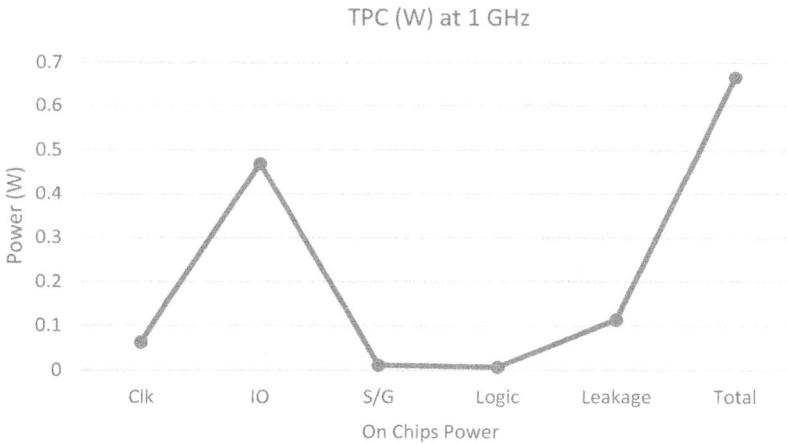

Figure 11.12 Representation of on-chips power for 1 GHz.

We know that the DP is the summation of clk, IO, S/G, and logic power. The clk, IO, S/G, and logic power are 0.013, 0.117, 0.002, and 0.001 W, respectively, for 1 GHz frequency. The graphical representation of on-chips power of the device is shown in Figure 11.12.

11.4.5 Power analysis for 5GHz frequency

As the frequency of operation is increased to the value of 5 GHz, the TP of the device measured is 0.666 W, which is the cumulative sum of SP (0.114 W); 17% of the TP and DP (0.552 W); and 83% of the TP. The DP is the total of clk, S/G, logic, and IO power, which are 0.063, 0.012, 0.007, and 0.470 W, respectively. The representation of the total on-chips power is shown in Figure 11.13, and the graphical representation is described in Figure 11.14.

On-Chip Power

	☐ Dynamic:	0.552 W (83%)
	■ Clocks:	0.063 W (11%)
	☐ Signals:	0.012 W (2%)
	■ Logic:	0.007 W (1%)
	☐ I/O:	0.470 W (86%)
	■ Device Static:	0.114 W (17%)

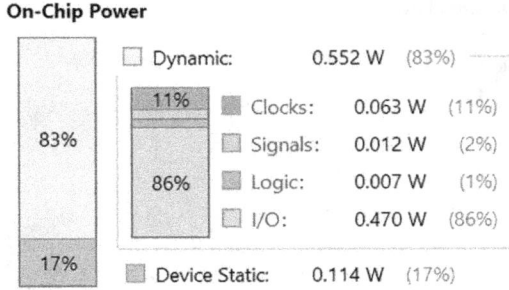

Figure 11.13 TP consumption for MOBILE DDR IO at 5 GHz frequency.

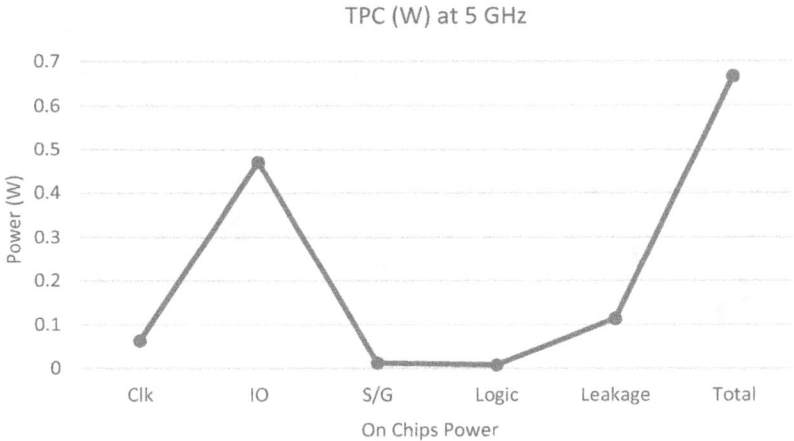

TPC (W) at 5 GHz

Figure 11.14 Representation of on-chips power for 5 GHz.

11.5 OBSERVATION AND ANALYSIS

As the frequency of operation for the device increases, the TPC for the MOBILE DDR IO increases, as shown in Section 11.4. The variation in TP is observed due to the variation in SP and DP of the FPGA device. The SP is constant for frequency from 100 MHz to 1 GHz. The SP changes only at 5 GHz frequency value. This variation in SP and DP causes the TPC to increase. Figure 11.15 depicts the variation in TPC for various frequency values.

TPC (W)

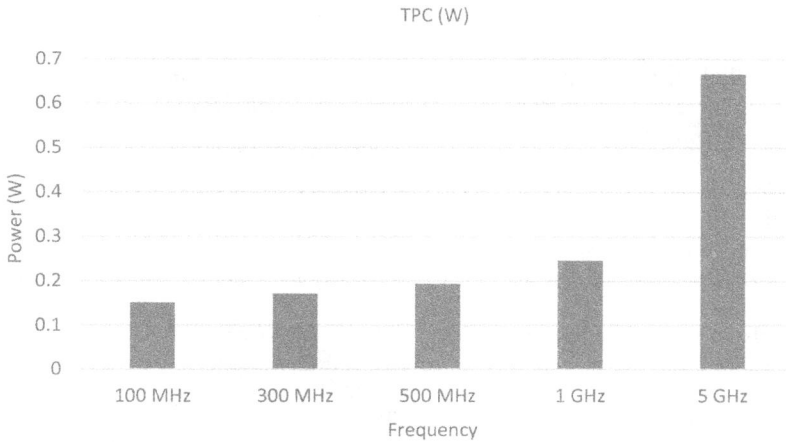

Figure 11.15 Variation in TPC for various frequency values.

As the frequency of operation for MOBILE DDR IO increases, the TPC of the device also increases, as shown in Figure 11.15. The optimal MOBILE DDR IO frequency for this device is 100 MHz.

11.6 CONCLUSION

It has been observed that over the past few years, the world has been experiencing a severe energy and power shortage. The longevity of the earth's natural resources is uncertain. Eventually, it must vanish. In the context of such a massive problem, the concept of GC comes to mind. This chapter is an advancement in the GCC field. Using the Kintex-7 device, a power-efficient FIR filter model is designed in this chapter. To optimize the power consumption of FIR filter, the authors of this chapter have employed the impedance matching technique, for which the MOBILE DDR IO standards have been used. In order to match the input and output impedance of the circuit, FPGA IO standards are utilized. This chapter examines the TPC of the device at various frequency values for the MOBILE DDR IO. The TPC of the device is observed to increase as the frequency of operation increases. The device provides the optimal amount of power at 100 MHz, while it consumes the maximum amount of power at 5 GHz.

In terms of its future applicability, the FIR filter design can be implemented with other SoC-based FPGAs, and various power optimization techniques, such as capacitance scaling, voltage/current variation of the device, and clock gating, can be used. In addition, the FPGA design can be converted into an ASIC design for enhanced performance.

REFERENCES

1. Koilpillai, R. David, and P. P. Vaidyanathan. "Cosine-modulated FIR filter banks satisfying perfect reconstruction." *IEEE Transactions on Signal Processing* 40, no. 4 (1992): 770–783.
2. Vaidyanathan, P., and Truong Nguyen. "Eigenfilters: A new approach to least-squares FIR filter design and applications including Nyquist filters." *IEEE Transactions on Circuits and Systems* 34, no. 1 (1987): 11–23.
3. Lim, Yong, and Sydney Parker. "FIR filter design over a discrete powers-of-two coefficient space." *IEEE Transactions on Acoustics, Speech, and Signal Processing* 31, no. 3 (1983): 583–591.
4. Hsiao, Shen-Fu, Jun-Hong Zhang Jian, and Ming-Chih Chen. "Low-cost FIR filter designs based on faithfully rounded truncated multiple constant multiplication/accumulation." *IEEE Transactions on Circuits and Systems II: Express Briefs* 60, no. 5 (2013): 287–291.
5. Davidson, Timothy N., Zhi-Quan Luo, and Jos F. Sturm. "Linear matrix inequality formulation of spectral mask constraints with applications to FIR filter design." *IEEE Transactions on Signal Processing* 50, no. 11 (2002): 2702–2715.
6. Hsiao, Shen-Fu, Jun-Hong Zhang Jian, and Ming-Chih Chen. "Low-cost FIR filter designs based on faithfully rounded truncated multiple constant multiplication/accumulation." *IEEE Transactions on Circuits and Systems II: Express Briefs* 60, no. 5 (2013): 287–291.
7. Vetterli, Martin, and Didier Le Gall. "Perfect reconstruction FIR filter banks: Some properties and factorizations." *IEEE Transactions on Acoustics, Speech, and Signal Processing* 37, no. 7 (1989): 1057–1071.
8. Mahapatra, Rajarshi, Yogesh Nijsure, Georges Kaddoum, Naveed Ul Hassan, and Chau Yuen. "Energy efficiency tradeoff mechanism towards wireless green communication: A survey." *IEEE Communications Surveys & Tutorials* 18, no. 1 (2015): 686–705.
9. Pandey, Bishwajeet, Bhagwan Das, Amanpreet Kaur, Tanesh Kumar, Abdul Moid Khan, D. M. Akbar Hussain, and Geetam Singh Tomar. "Performance evaluation of FIR filter after implementation on different FPGA and SOC and its utilization in communication and network." *Wireless Personal Communications* 95, no. 2 (2017): 375–389.
10. Kumar, Tanesh, Bishwajeet Pandey, Teerath Das, and B. S. Chowdhry. "Mobile DDR IO standard based high performance energy efficient portable ALU design on FPGA." *Wireless Personal Communications* 76, no. 3 (2014): 569–578.
11. Kintex-7 FPGA family. https://www.xilinx.com/products/silicon-devices/fpga/kintex-7.html. Accessed on 30 June 2023.
12. Bhat, Deepshikha, Amanpreet Kaur, and Sunny Singh. "Wireless sensor network specific low power FIR filter design and implementation on FPGA." In *2015 2nd International Conference on Computing for Sustainable Global Development (INDIACom)*, India, pp. 1534–1536. IEEE, 2015.
13. Pandey, Bishwajeet, Nisha Pandey, Amanpreet Kaur, D. M. Akbar Hussain, Bhagwan Das, and Geetam S. Tomar. "Scaling of output load in energy efficient FIR filter for green communication on ultra-scale FPGA." *Wireless Personal Communications* 106, no. 4 (2019): 1813–1826.

14. Kaur, Ravinder, Jagdish Kumar, Sumita Nagah, Bishwajeet Pandey, and Kavita Goswami. "IO Standard based low power memory design and implementation on FPGA." In *2015 2nd International Conference on Computing for Sustainable Global Development (INDIACom)*, India, pp. 1501–1505. IEEE, 2015.
15. Sharma, Rashmi, Bishwajeet Pandey, Vikas Jha, Siddharth Saurabh, and Sweety Dabas. "Input–output standard-based energy efficient UART design on 90nm FPGA." In Sunil Kumar Muttoo (Ed.), *System and architecture*, pp. 139–150. Springer, Singapore, 2018.

Chapter 12

LVCMOS-based packet counter for GCC

LIST OF ABBREVIATIONS

AC	Alternating Current
ASIC	Application-Specific Integrated Circuit
BUFG	Global Buffers
Clk	Clocks
CMOS	Complementary Metal-Oxide-Semiconductor
DDR	Double Data Rate
DRAM	Dynamic Random-Access Memory
DP	Dynamic Power
FF	Flip-Flops
FPGA	Field Programmable Gate Array
GC	Green Computing
GCC	Green Communication Computing
G. Comm.	Green Communication
GPS	Global Positioning System
IC	Integrated Circuit
IO	Input Output
IT	Information Technology
JT	Junction Temperature
LP	Leakage Power
LUT	Look Up Tables
LVCMOS	Low-Voltage Complementary Metal-Oxide-Semiconductor
RTL	Register Transfer Logic
S/G	Signal
SP	Static Power
TCP/IP	Transmission Control Protocol/Internet Protocol
TM	Thermal Margin
TP	Total Power
TPC	Total Power Consumption
ϑJA	Effective Thermal Resistance to Air

DOI: 10.1201/9781003302872-12

12.1 INTRODUCTION

In the age of information technology (IT), it seems quite easy to transfer the data from one node to another node or one device to the other device very efficiently. The data may be transferred within a few seconds. The data being transferred from one device to the other happens in the form of packets. Such packets are also known as data packets or simply packets. The transfer of packets happens only by the use of networking concepts. When the data is transferred over the Internet, there might be some chance of packets being missed or delayed in reaching its destination [1,2]. Therefore, in the age of IT, any missing or delay in such kind of data may lead to some critical issues for some organization. Such kinds of problems occurring in the data transmission can be rectified using the "packet counter." The "packet counter" is software for confirming the data transmission size and the approximate data communications charges of PacketWIN [3,4]. It monitors the data packets which are transferred from the source to the destination.

Computer networking is the term that is used in transferring the data packet from the source to the destination. Transmission of data packets is done through the different layers of TCP/IP layers [5,6]. There are generally five layers of the TCP/IP model as shown in Figure 12.1.

The packets counter monitors and counts the each and every packet sent from the physical layer to the application layer. The packet counter is responsible for the successful transfer of data packets from physical layer to application layer [7,8]. There are several features of packet counter which are listed below:

a. Communications charge: The data communications size and the communications charge according to the selected rate can be presented in real time or after the communication, respectively.
b. Rate configuration: The customer's contracted rate can be entered in advance, and communication charges can be presented based on this rate.

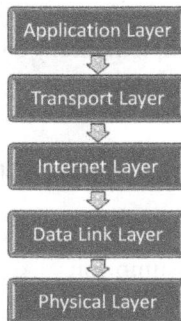

Figure 12.1 TCP/IP layer.

c. Data size monitoring: A warning window can be presented when the data size exceeds the predetermined number of bytes (number of packets).

d. The communications log: One can save a conversation log in CSV format (start time, end time, communications volume, communications charges, and communications destination address).

The packet counter might consume huge amount of power at the time of operation. Therefore, we should focus on such a system or a model which ensures the concepts and promotes the idea of GC, which means designing a power-efficient packet counter [9,10]. This chapter highlights about the design of a power-efficient model of packet counter which will not only ensure GC but will also help in efficient transfer of data packets. The power-efficient model of packet counter is designed using the FPGA device. FPGAs are those semiconductor devices which are used for implementation of digital circuits. The main advantage of the FPGA device is that it can be reconfigured after its manufacturing [11,12]. Therefore, in this chapter, we will discuss how the packet counter can be made power-efficient using the FPGA device.

12.2 IMPLEMENTATION OF PACKET COUNTER ON FPGA

For the development of a power-efficient packet counter model, the Kintex-7 FPGA device was utilized. It is a Xilinx family device with a 28 nm gate size. Using the VIVADO ISE, the packet counter design is implemented on the same device. To optimize power consumption, LVCMOS IOs are employed. IO standard matches the transmission line impedance for the internal circuit in order to minimize power consumption. To implement the design on the FPGA device, FPGA resources such as IO, FF, LUTs, and BUFG are utilized [13,14]. The resource utilization for packet counter design is displayed in Table 12.1 and illustrated in Figure 12.2.

From Table 12.1, it is observed that in the process of designing the packet counter design 71 IOs and 4 LUTs are consumed as well as 38 FFs and

Table 12.1 Resource utilization for designing packet counter design

Resources	Available	Utilization	Utilization (%)
LUT	4	41,000	0.01
IO	71	300	23.67
BUFG	1	32	3.13
FF	38	82,000	0.05

RESOURCE UTILIZATION

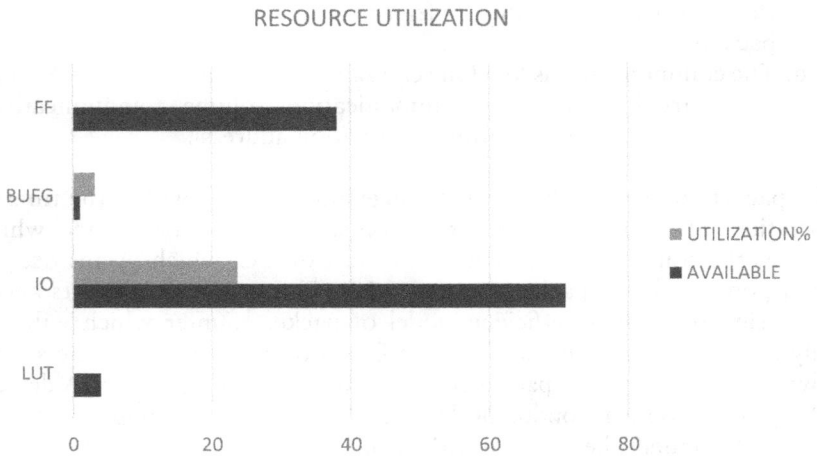

Figure 12.2 Resource utilization for designing packet counter design.

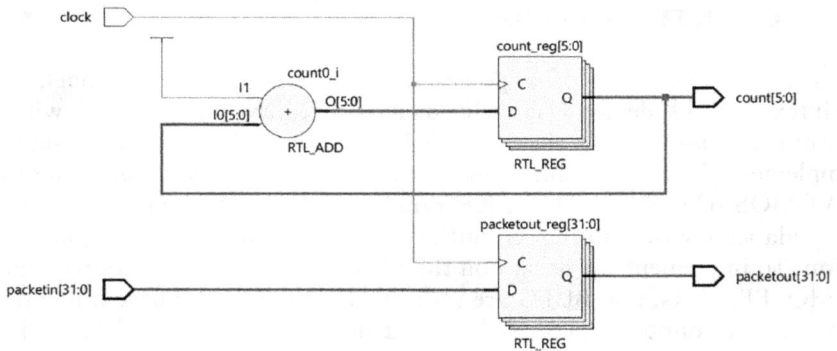

Figure 12.3 RTL schematic of packet counter.

1 BUFG of Kintex-7 device. The RTL design of packet counter is shown in Figure 12.3, while the technology schematic of the packet counter is described in Figure 12.4 [15,16].

Figure 12.3 describes the RTL of the packet counter. From Figure 12.3, it is observed that there are two blocks of counter which counts the bits. At the input end, we have a clk pulse generator, along with a 32-bit packet input. At the output end, the counter counts the bits according to the clk pulse and gives the packet outputs (32-bits).

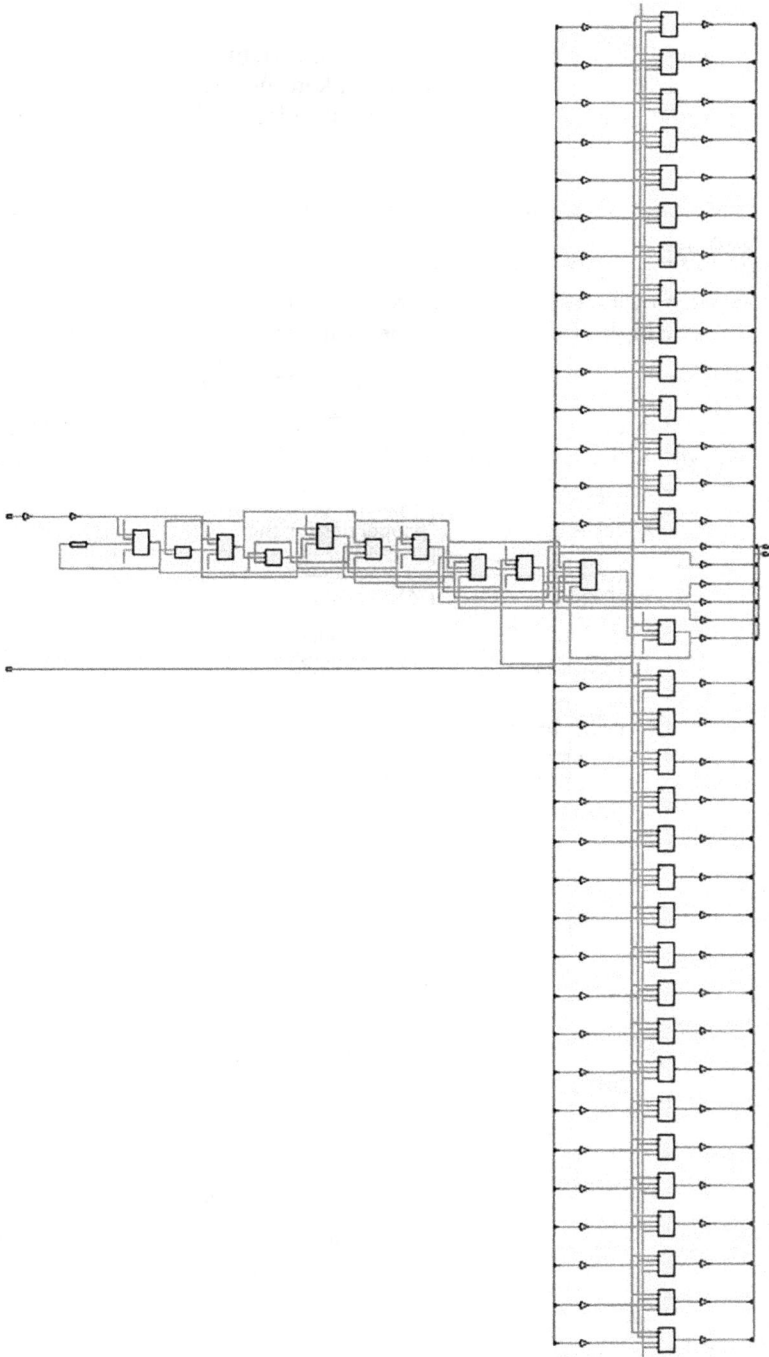

Figure 12.4 Technology schematic of packet counter.

12.3 THERMAL PROPERTIES

This section will give a brief understanding about the thermal properties of the device. The thermal properties are taken into account while matching the impedance using the LVCMOS IO standard. The thermal properties considered here are as follows:

a. TM: This is a device's attribute that permits it to consume very little power.
b. ϑJA: When 1 W of power is dissipated in the IC, the amount of heat created or a rise in temperature is specified.
c. JT: The highest operating temperature of an FPGA device's integrated gates.

The thermal properties for the Kintex-7 device, when the impedance is matched with LVCMOS IO, are well described in Table 12.2 and Figure 12.5.

Table 12.2 Thermal properties for the LVCMOS IO for SPRATAN-7

IO std	JT (°C)	TM (°C)	ϑJA (°C/W)
LVCMOS 12	26.1	58.9	1.9
LVCMOS 15	42.2	42.8	1.9
LVCMOS 18	47.8	37.2	1.9
LVCMOS 25	64.5	20.5	1.9
LVCMOS 33	92.5	−7.5	1.9

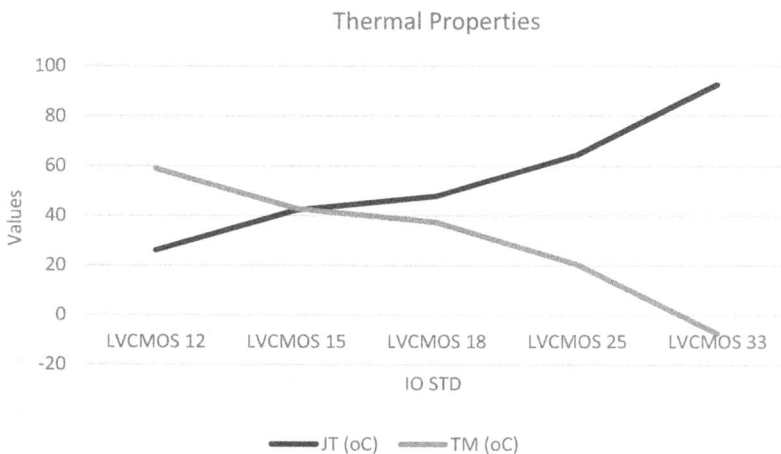

Figure 12.5 Thermal properties for the LVCMOS IO for Kintex-7.

Table 12.2 demonstrates that as the input voltage of transmission line impedance matching increases, so does the JT. For LVCMOS IO, the TM decreases as the input voltage increases. The TM value for the LVCMOS 33 IO becomes negative. Additionally, for the LVCMOS 33, the junction gate of the internally connected MOSFET is destroyed (red color). ϑJA remains constant at 1.9 across all LVCMOS IO.

12.4 POWER ANALYSIS

In the present state of the globe, the power consumption of gadgets and information technology poses the greatest threat. The whole globe is experiencing an energy deficit. Thus, the whole ecosystem collaborates to produce a low-power solution for the electronics and information technology industries. Reduced power consumption prolongs the device's life. In this part, we calculate the power consumption of a packet counter implementation on a Kintex-7 device using the LVCMOS IO standard. The device's total power consumption is the sum of both dynamic and static power consumption [17,18]. This is described mathematically in Equation 12.1:

$$TP = SP + DP \qquad (12.1)$$

where

TP = Total Power
SP = Static Power
DP = Dynamic Power

12.4.1 Power analysis for LVCMOS 12

When the impedance is matched with the LVCMOS IO, the total power consumption (TPC) observed for the device is 0.598 W. The DPC is 87% of

On-Chip Power

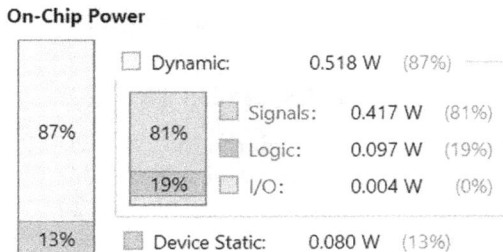

	Dynamic:	0.518 W	(87%)
87%	81%	Signals:	0.417 W	(81%)
		Logic:	0.097 W	(19%)
	19%	I/O:	0.004 W	(0%)
13%	Device Static:	0.080 W	(13%)

Figure 12.6 On-chips power consumption for LVCMOS 12 IO.

TPC (W) for LVCMOS 12 IO

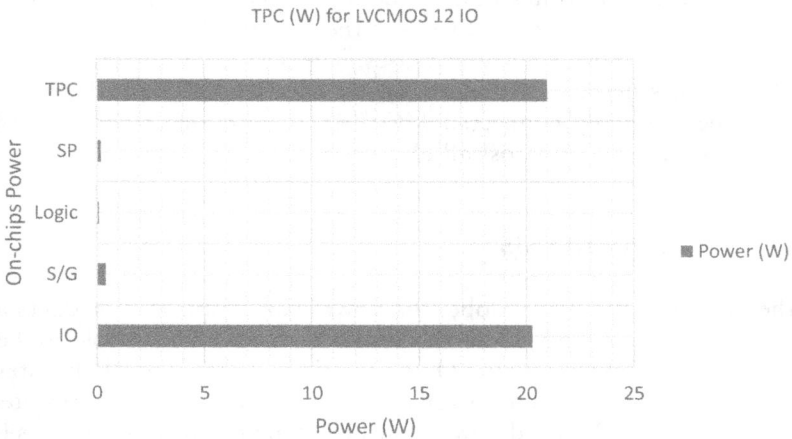

Figure 12.7 Graphical representation of the TPC for LVCMOS 12.

the TPC which is 0.518 W, and SPC is 13% of the TPC which is 0.080 W. The on-chips power consumption for the device when impedance is matched with LVCMOS 12 IO is described in Figure 12.6.

From Figure 12.6, it can be clearly observed that the DPC is the combination of signal (S/G), logic (L/G), and IO power of the device. These are 0.417, 0.097, and 0.004 W, respectively. The graphical representation of the TPC for LVCMOS 12 is described in Figure 12.7.

12.4.2 Power analysis for LVCMOS 15

When the LVCMOS 15 IO is used for matching the impedance of the transmission line, it is observed that the DPC is 9.032 W, which is 99% of the TPC of the device. The SPC is just 1% of the TPC, which is 0.105 W. The TPC is the summation of SPC and DPC which is 9.137 W. The on-chips power consumption is described in Figure 12.8.

The DPC is the combination of S/G, L/G, and IO power of the device which are 0.417, 0.097, 8.518 W, respectively. The graphical representation of the TPC for LVCMOS 15 is presented in Figure 12.9.

12.4.3 Power analysis for LVCMOS 18

When the impedance is matched with the LVCMOS 18 IO, the total power consumption (TPC) observed for the device is 12.106 W. The DPC is 99% of the TPC which is 11.986 W, and SPC is 1% of the TPC which is 0.119 W. The on-chips power consumption for the device when impedance is matched with LVCMOS 18 IO is described in Figure 12.10.

On-Chip Power

☐ Dynamic:	9.032 W	(99%)	
▦ Signals:	0.417 W	(5%)	
▦ Logic:	0.097 W	(1%)	
☐ I/O:	8.518 W	(94%)	
▦ Device Static:	0.105 W	(1%)	

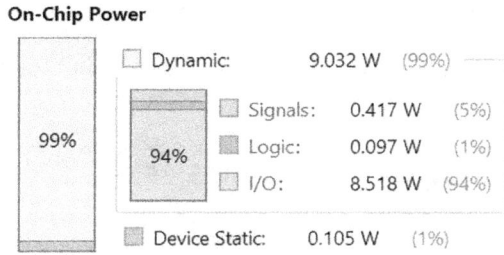

Figure 12.8 On-chips power consumption for LVCMOS 15.

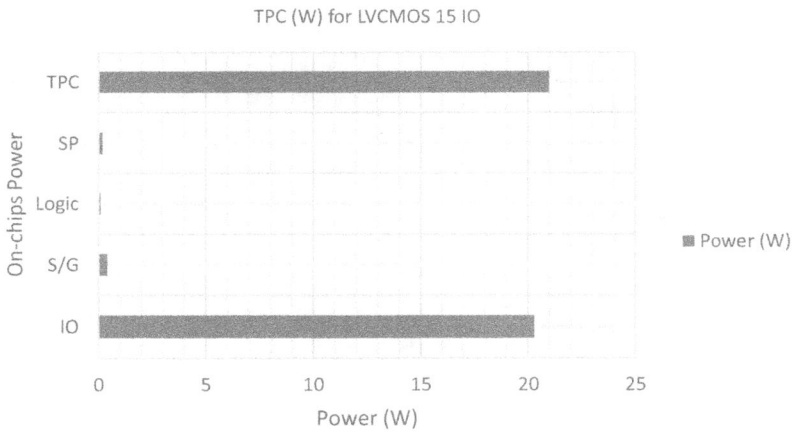

TPC (W) for LVCMOS 15 IO

Figure 12.9 Graphical representation of the TPC for LVCMOS 15.

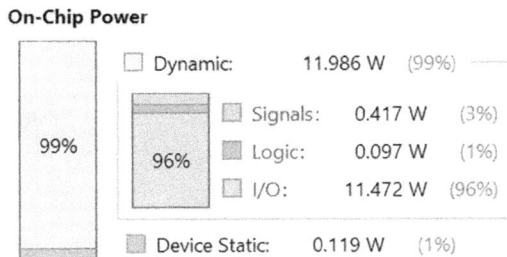

On-Chip Power

☐ Dynamic:	11.986 W	(99%)	
▦ Signals:	0.417 W	(3%)	
▦ Logic:	0.097 W	(1%)	
☐ I/O:	11.472 W	(96%)	
▦ Device Static:	0.119 W	(1%)	

Figure 12.10 On-chips power consumption for LVCMOS 18 IO.

From 12.10, it can be clearly observed that the DPC is the combination of S/G, L/G, and IO power of the device. These are 0.417, 0.097, and 11.472 W, respectively. The graphical representation of the TPC for LVCMOS 18 is described in Figure 12.11.

12.4.4 Power analysis for LVCMOS 25

When the LVCMOS 25 IO is used for matching the impedance of the transmission line, it is observed that the DPC is 20.804 W, which is 99% of the TPC of the device. The SPC is just 1% of the TPC, which is 0.182 W. The TPC is the summation of SPC and DPC which is 20.986 W. The on-chips power consumption is described in Figure 12.12.

TPC (W) for LVCMOS 18 IO

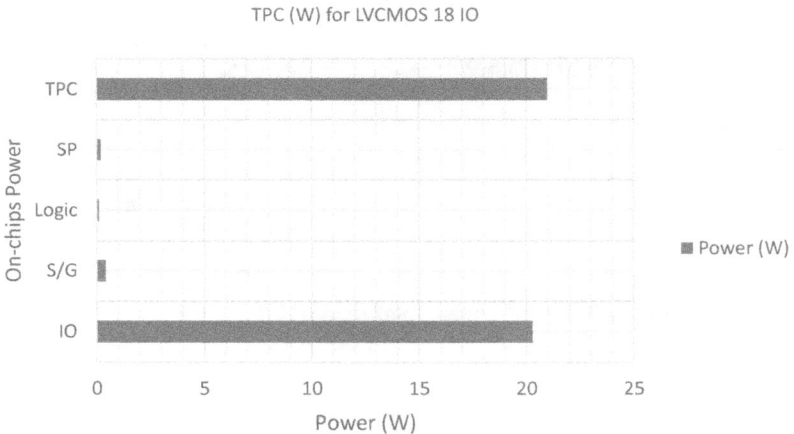

Figure 12.11 Graphical representation of the TPC for LVCMOS 18.

On-Chip Power

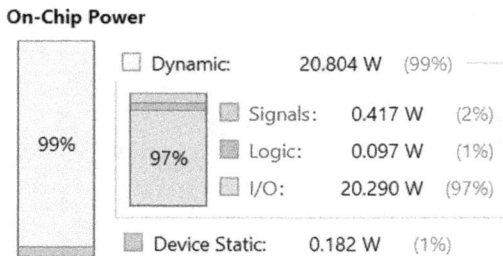

☐ Dynamic:	20.804 W	(99%)
▩ Signals:	0.417 W	(2%)
▩ Logic:	0.097 W	(1%)
☐ I/O:	20.290 W	(97%)
▩ Device Static:	0.182 W	(1%)

Figure 12.12 On-chips power consumption for LVCMOS 25.

The DPC is the combination of S/G, L/G, and IO power of the device which are 0.417, 0.097, 20.290 W, respectively. The graphical representation of the TPC for LVCMOS 25 is presented in Figure 12.13.

12.4.5 Power analysis for LVCMOS 33

When the impedance is matched with LVCMOS 33 IO, it is observed that the JT of the device gets exceeded and hence the junction gate of the device gets burned. The TPC of the devices calculated is 35.852 W. The on-chips power consumption of the device is represented in Figure 12.14.

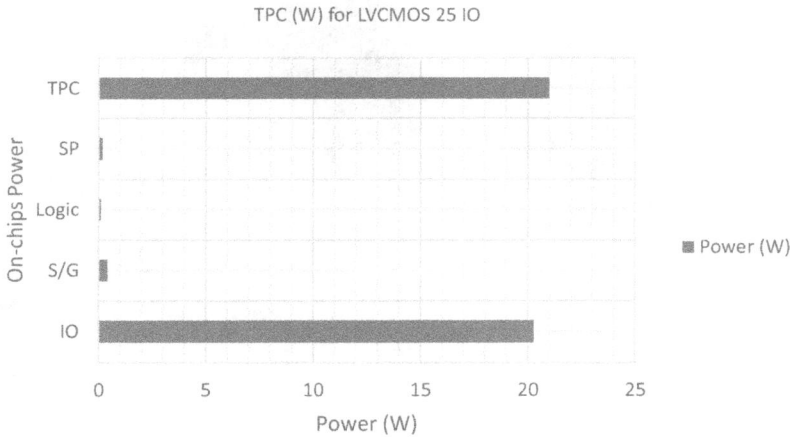

Figure 12.13 Graphical representation of the TPC for LVCMOS 25.

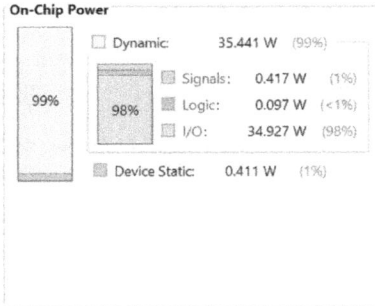

Power analysis from Implemented netlist. Activity derived from constraints files, simulation files or vectorless analysis.

Total On-Chip Power:	35.852 W (Junction temp exceeded!)
Design Power Budget:	**Not Specified**
Power Budget Margin:	**N/A**
Junction Temperature:	92.5°C
Thermal Margin:	-7.5°C (-3.9 W)
Effective θJA:	1.9°C/W
Power supplied to off-chip devices:	0 W
Confidence level:	Low

Launch Power Constraint Advisor to find and fix invalid switching activity

On-Chip Power

	Dynamic:	35.441 W	(99%)
	Signals:	0.417 W	(1%)
	Logic:	0.097 W	(<1%)
	I/O:	34.927 W	(98%)
	Device Static:	0.411 W	(1%)

99%   98%

Figure 12.14 On-chips power for LVCMOS 33 IO.

12.5 OBSERVATION AND ANALYSIS

This section will give insights about the analysis and observation observed from Section 12.4. From Section 12.4, it is observed that the TPC increases as the input voltage of the transmission line increases; i.e., TPC increases for LVCMOS IO as shown in Figure 12.15.

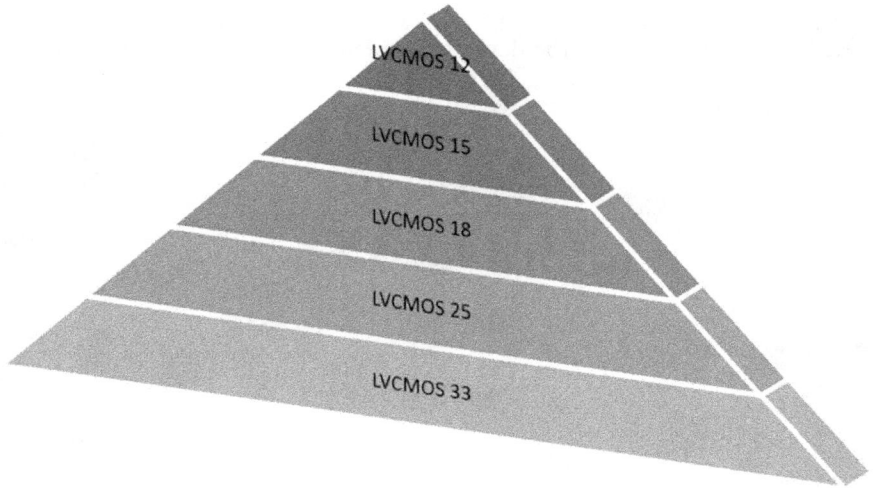

Figure 12.15 TPC increasing for LVCMOS IO.

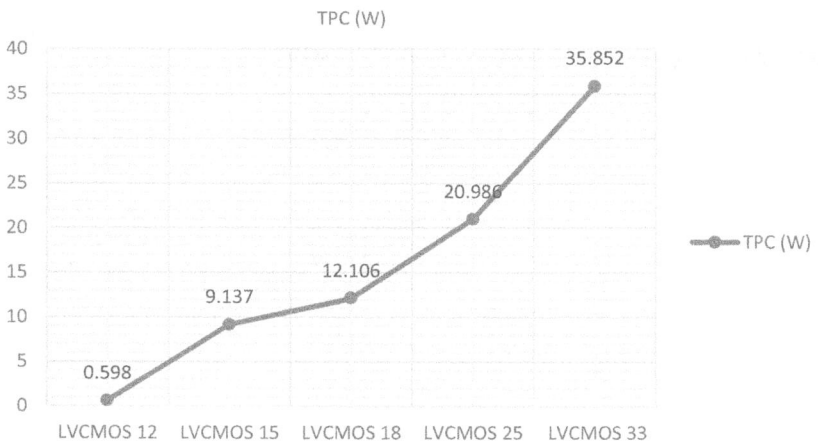

Figure 12.16 TPC of the device for LVCMOS family.

From Section 12.4, it is observed that TPC is minimum for LVCMOS 12 IO, and the TPC is maximum for LVCMOS 33 IO. Also, when the impedance is matched with LVCMOS 33 IO, the JT gets exceeded and hence the junction gate gets burned, and so the device fails to work. The TPC of the device for LVCMOS family IO is described in Figure 12.16.

12.6 CONCLUSION

This chapter gives a detailed view of the TPC of the packet counter on the KINTEX-7 FPGA. The packet counter is one of the key components used in the world of signal processing; apart from this, it has also several DSP applications. In this chapter, the power consumption of the packet counter is analyzed over KINTEX-7. The design of packet counter is implemented on VIVADO ISE. The power consumption is tried to be optimized with LVCMOS IO. IO standards are used to match the impedance of the input line with output line impedance. The power consumption of any device is optimized when there is a perfect impedance matching. Here various LVCMOS IOs are used to match the impedance. From Section 12.4, it is observed that as the input voltage of the IO standard increases, the TPC also increases. Therefore, the device gives the optimal power consumption when the input voltage is low. The packet counter design on KINTEX-7 devices has the optimal power consumption when the impedance is matched with LVCMOS 12 IO. As far as future scope is concerned, the packet counter design can be implemented over several other FPGAs of distinguished family such as Artix-7, Spartan-7, and Virtex-7. Also, the designs can be converted into ASIC design for better applications and uses.

REFERENCES

1. Whiting, Doug, Russ Housley, and Niels Ferguson. Counter with CBC-MAC (CCM). No. rfc3610. 2003.
2. Housley, Russell. Using advanced encryption standard (AES) counter mode with IPsec encapsulating security payload (ESP). No. rfc3686. 2004.
3. Packet counter I Software download. au. https://www.au.com/english/mobile/service/mobile-communications/soft-download/packet-counter/. Accessed on 30 June 2023.
4. Yang, Jianwei, Fan Dai, Jielin Wang, Jianmin Zeng, Zhang Zhang, Jun Han, and Xiaoyang Zeng. "Countering power analysis attacks by exploiting characteristics of multicore processors." *IEICE Electronics Express* 15, no. 7 (2018): 20180084.
5. Kaur, Damandeep, and Parminder Singh. "Various OSI layer attacks and countermeasure to enhance the performance of WSNs during wormhole attack." *International Journal on Network Security* 5, no. 1 (2014): 62.

6. Klein, Randall W., Michael A. Temple, and Michael J. Mendenhall. "Application of wavelet-based RF fingerprinting to enhance wireless network security." *Journal of Communications and Networks* 11, no. 6 (2009): 544–555.

7. Sinha, Preeti, V. K. Jha, Amit Kumar Rai, and Bharat Bhushan. "Security vulnerabilities, attacks and countermeasures in wireless sensor networks at various layers of OSI reference model: A survey." In *2017 International Conference on Signal Processing and Communication (ICSPC)*, Limassol, Cyprus, pp. 288–293. IEEE, 2017.

8. Banerjee, Suman, Bobby Bhattacharjee, and Christopher Kommareddy. "Scalable application layer multicast." In *Proceedings of the 2002 Conference on Applications, Technologies, Architectures, and Protocols for Computer Communications*, Pittsburg, PA, pp. 205–217. ACM Press, 2002.

9. Gupta, Isha, Swati Singh Garima, Harpreet Kaur, Deepshikha Bhatt, and Aamir Vohra. "28nm FPGA based power optimized UART design using HSTL I/O standards." *Indian Journal of Science and Technology* 8, no. 17 (2015): 1–6.

10. Kumar, Keshav, Amanpreet Kaur, S. N. Panda, and Bishwajeet Pandey. "Effect of different nano meter technology-based FPGA on energy efficient UART design." In *2018 8th International Conference on Communication Systems and Network Technologies (CSNT)*, India, pp. 1–4. IEEE, 2018.

11. Kumar, Keshav, Amanpreet Kaur, Bishwajeet Pandey, and S. N. Panda. "Low power UART design using different nanometer technology-based FPGA." In *2018 8th International Conference on Communication Systems and Network Technologies (CSNT)*, India, pp. 1–3. IEEE, 2018.

12. Kumar, Keshav, Bishwajeet Pandey, Amit Kant Pandit, Y. A. Baker El-Ebiary, Salameh A. Mjlae, and Samer Bamansoor. "Design of low power transceiver on spartan-3 and spartan-6 FPGA." *International Journal of Innovative Technology and Exploring Engineering* 8, no. 12S2 (2019): 27–30.

13. Sandhu, Amanpreet, Vidhoytma Gandhi, Simranpreet Kaur, Surbhi Huria, Divjot Singh, and Wamika Goyal. "Thermally aware LVCMOS based low power universal asynchronous receiver transmitter design on FPGA." *Indian Journal of Science and Technology* 8, no. 20 (2015): 1–4.

14. Kumar, Abhishek, Bishwajeet Pandey, D. M. Akbar Hussain, Mohammad Atiqur Rahman, Vishal Jain, and Ayoub Bahanasse. "Low voltage complementary metal oxide semiconductor-based energy efficient UART design on Spartan-6 FPGA." In *2019 11th International Conference on Computational Intelligence and Communication Networks (CICN)*, India, pp. 84–87. IEEE, 2019.

15. Pandey, Bishwajeet, and Ravikant Kumar. "Low voltage DCI based low power VLSI circuit implementation on FPGA." In *2013 IEEE Conference on Information & Communication Technologies*, India, pp. 128–131. IEEE, 2013.

16. Kaur, Harkinder, Harsh Sohal, and Jaiteg Singh. "Design and performance analysis of UART using Altera Quartus-II and Xilinx ISE 14.2." In *6th International Conference on Communication and Network Technologies*, India. 2016.

17. Kaur, Ravinder, Jagdish Kumar, Sumita Nagah, Bishwajeet Pandey, and Kavita Goswami. "IO Standard based low power memory design and implementation on FPGA." In *2015 2nd International Conference on Computing for Sustainable Global Development (INDIACom)*, India, pp. 1501–1505. IEEE, 2015.

18. Kumar, Vivek, Aksh Rastogi, and V. K. Tomar. "Implementation of UART design for RF modules using different FPGA technologies." *IOP Conference Series: Materials Science and Engineering* 1116, no. 1 (2021): 012131.

Chapter 13

SSTL-based packet counter of GCC

LIST OF ABBREVIATIONS

AC	Alternating Current
ASIC	Application-Specific Integrated Circuit
BUFG	Global Buffers
Clk	Clocks
CMOS	Complementary Metal-Oxide-Semiconductor
DDR	Double Data Rate
DRAM	Dynamic Random-Access Memory
DP	Dynamic Power
EB	Exabytes
FF	Flip-Flops
FPGA	Field Programmable Gate Array
GC	Green Computing
GCC	Green Communication Computing
G. Comm.	Green Communication
GPS	Global Positioning System
IC	Integrated Circuit
IO	Input Output
IT	Information Technology
JT	Junction Temperature
LP	Leakage Power
LUT	Look Up Tables
LVCMOS	Low-Voltage Complementary Metal-Oxide-Semiconductor
RTL	Register Transfer Logic
S/G	Signal
SP	Static Power
SSTL	Stub Series Terminated Logic
TCP/IP	Transmission Control Protocol/Internet Protocol
TM	Thermal Margin
TP	Total Power
TPC	Total Power Consumption
ϑJA	Effective Thermal Resistance to Air

DOI: 10.1201/9781003302872-13

13.1 INTRODUCTION

The present age is the IT age (also known as the Digital Age, Silicon Age, Computer Age, or New Media Age). It started in the middle of the twentieth century and is characterized by a fast transition from conventional industries to an IT-based economy. The world's technical storage capacity expanded from 2.6 (optimally compressed) exabytes (EB) in 1986 to 15.8 (EB) in 1993, to 54.5 (EB) in 2000, and 295 (EB) in 2007. It is projected that the world's information storage capacity hit 5 zettabytes in 2014, which is the equal of 4,500 books piled from the earth to the sun [1].

Similar to Moore's law, it seems that the quantity of digital data saved is rising roughly exponentially. According to Kryder's law, the quantity of storage capacity accessible looks to be growing nearly exponentially. In 1986, the technological capacity of the world to receive information via one-way broadcast networks was 432 exabytes of (optimally compressed) information; in 1993, it was 715 exabytes; in 2000, it was 1.2 zettabytes; and in 2007, it was 1.9 zettabytes, the information equivalent of 174 newspapers per person per day [2].

The world's effective capacity to communicate information over two-way telecommunication networks increased from 281 petabytes of (optimally compressed) data in 1986 to 471 petabytes in 1993 to 2.2 exabytes in 2000 to 65 exabytes per person in 2007. In the 1990s, the rapid spread of the Internet increased firms' and families' access to and capacity for sharing information. The increasing data volume necessitates the correct delivery of data packets for effective communication. If there are any missing or altered data packets, communication will break down [3]. Using a packet counter helps regulate the correct flow of packets from source to recipient. When data is exchanged through software, the packet counter functions as a software program and keeps track of each and every packet transported from sender to recipient. When data transport is accomplished through hardware, the packet counter software is built alongside a hardware microcontroller's counter circuit in order to determine the packet count [4].

In the world of internet, the data packets transferred from sender to receiver use the TCP/IP protocols. There are basically five different layers in the TCP/IP model which is described in Figure 13.1.

Each and every packet transmitted from the physical layer to the application layer is monitored and counted by the packets counter. The application layer is where communication occurs. The packet counter facilitates the successful transport of data packets from the physical layer to the application layer [5,6]. In the hardware implementation of the packet counter, it might be possible that the device may consume huge amount of power. Huge power consumption requires huge amount of utilization of resources, and hence the lifetime of the device is also be affected [7,8]. To ensure the minimum utilization of natural resources and for the betterment of the life

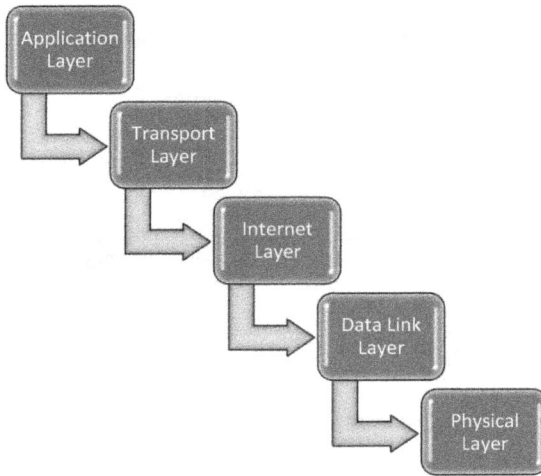

Figure 13.1 TCP/IP layer.

process of the device, one should focus on the ideas of GC. Moreover, GC also leads to sustainable development.

This chapter focuses on the creation of a power-efficient packet counter paradigm that not only ensures GC but also facilitates efficient data packet transmission. Using the FPGA device, a power-efficient type of packet counter is created. FPGAs are semiconductor devices that are used to build digital circuits. After manufacture, the FPGA device may be modified, which is its primary benefit [9–11]. For making it suitable for GC, we are using Kintex-7 FPGA with SSTL IO to match the impedance of the circuitry. In this chapter, we will explore how the FPGA device may be used to make the packet counter power-efficient.

13.2 IMPLEMENTATION OF PACKET COUNTER ON FPGA

For the development of a power-efficient packet counter model, the Kintex-7 FPGA device was utilized. It is a Xilinx family device with a 28 nm gate size. Using the VIVADO ISE, the packet counter design is implemented on the same device. To optimize power consumption, SSTL IOs are employed. IO standard matches the transmission line impedance for the internal circuit in order to minimize power consumption. To implement the design on the FPGA device, FPGA resources such as IO, FF, LUTs, and BUFG are utilized. The resource utilization for PACKET COUNTER design is displayed in Table 13.1 and illustrated in Figure 13.2.

From Table 13.1, it is observed that in the process of designing the packet counter design, 71 IOs and 4 LUTs are consumed as well as 38 FFs and 1 BUFG of Kintex-7 device. The RTL design of packet counter is shown in Figure 13.3, while the technology schematic of the packet counter is described in Figure 13.4.

Table 13.1 Resource utilization for designing PACKET COUNTER design

Resources	Available	Utilization	Utilization (%)
LUT	4	41,000	0.01
IO	71	300	23.67
BUFG	1	32	3.13
FF	38	82,000	0.05

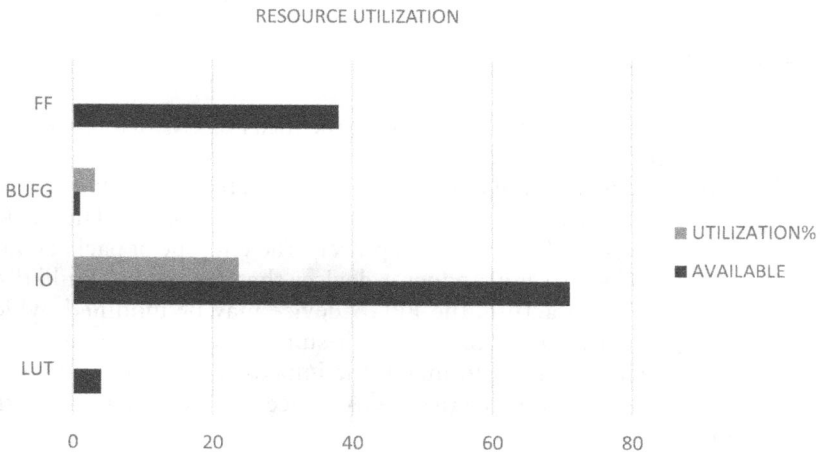

Figure 13.2 Resource utilization for designing PACKET COUNTER design.

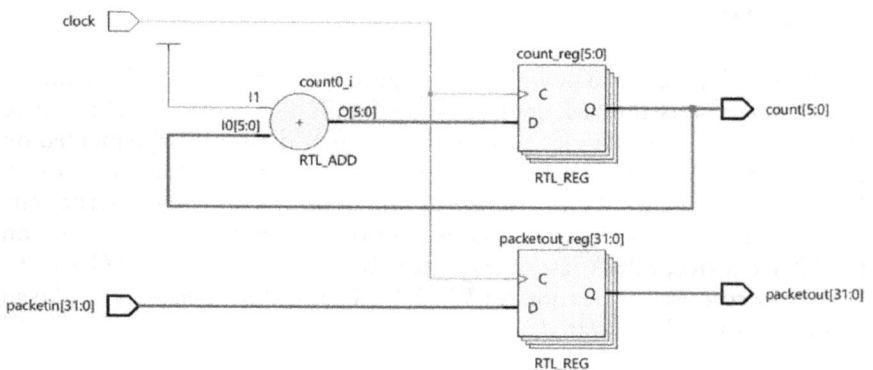

Figure 13.3 RTL schematic of packet counter.

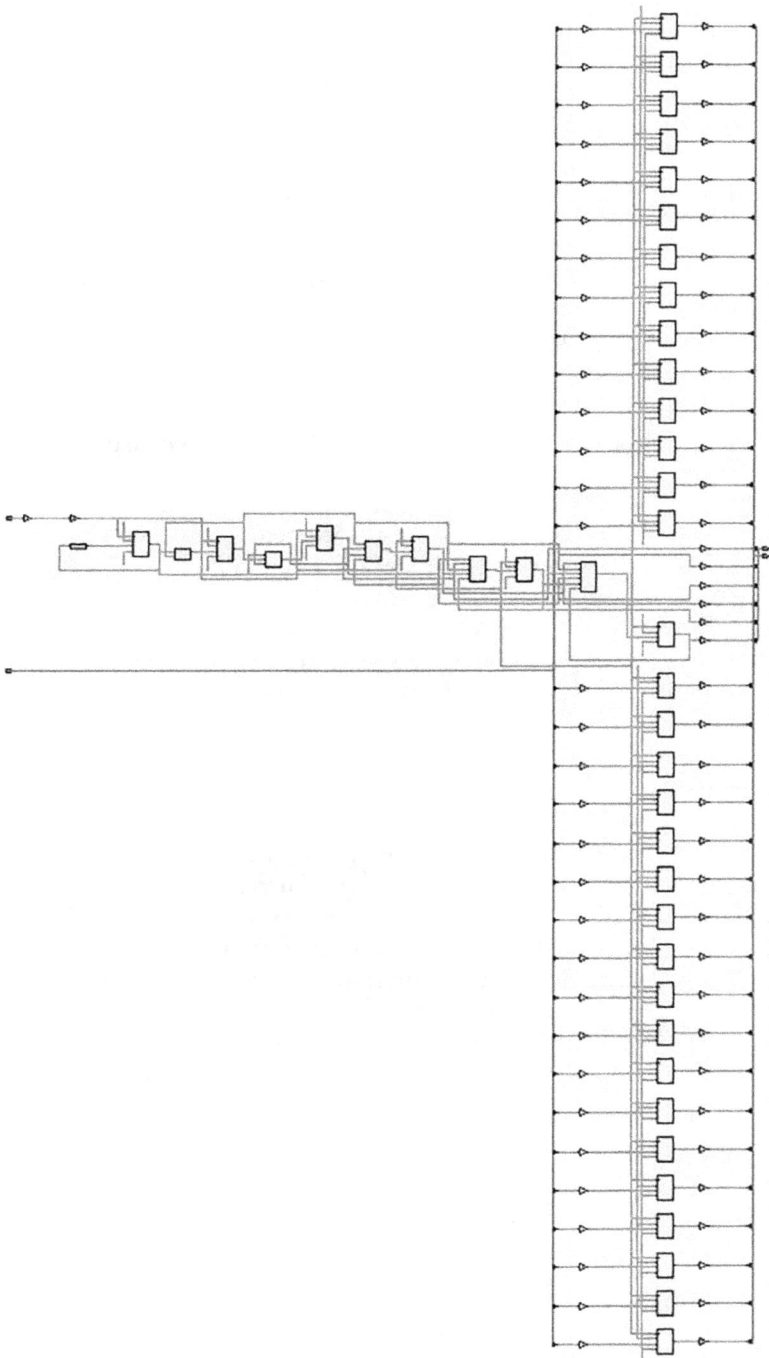

Figure 13.4 Technology schematic of packet counter.

Figure 13.3 describes the RTL of the packet counter. In Figure 13.3, it is observed that there are two blocks counter which counts the bits. At the input end, we have a clk pulse generator, along with a 32-bit packet input. At the output end, the counter counts the bits according to the clk pulse and gives the packet outputs (32-bits).

13.3 THERMAL PROPERTIES

This section will give a brief understanding about the thermal properties of the device. The thermal properties are taken in account while matching the impedance using the SSTL IO standard [12–14]. The thermal properties considered here are as follows:

a. TM: This is a device's attribute that permits it to consume very little power.
b. ϑJA: When 1 W of power is dissipated in the IC, the amount of heat created or a rise in temperature is specified.
c. JT: The highest operating temperature of an FPGA device's integrated gates.

The thermal properties for the Kintex-7 device, when the impedance is matched with SSTL IO, are well described in Table 13.2 and Figure 13.5.

13.4 POWER ANALYSIS

The world's greatest threat is becoming the energy consumption of electronic gadgets and information technology. Currently, the globe is suffering an energy scarcity. In consequence, the whole ecosystem collaborates to develop a low-power solution for the information technology and electronics sectors. Reduced power consumption increases the device's durability. In this part, we calculate the power consumption of a FIR implementation on a Kintex-7 device using the SSTL IO standard. The device's total power consumption is the sum of both dynamic and static power consumption [15–17]. This is described mathematically by Equation 13.1:

$$TP = SP + DP \tag{13.1}$$

where

TP = Total Power
SP = Static Power
DP = Dynamic Power

Table 13.2 Thermal properties for the SSTL IO for SPRATAN-7

IO std	JT (°C)	TM (°C)	θJA (°C/W)
SSTL 12	26.1	73.9	1.9
SSTL 135	39.6	60.4	1.9
SSTL 15	40.0	60.0	1.9
SSTL 18_I	49.0	51.0	1.9
SSTL 18_II	53.1	46.9	1.9

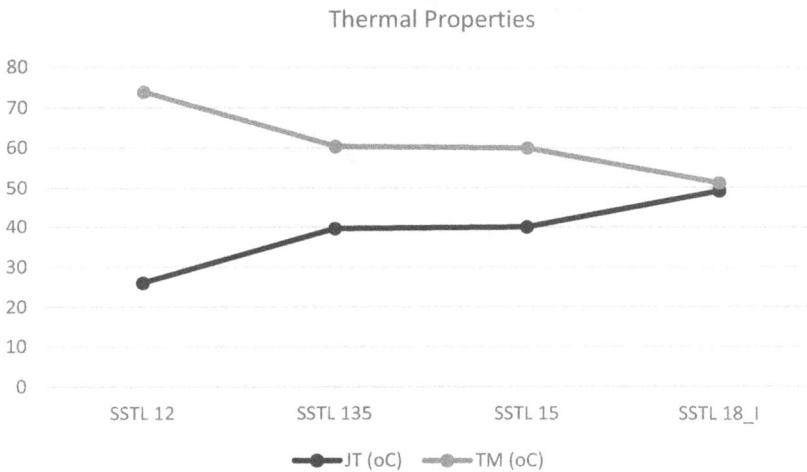

Figure 13.5 Thermal properties for the SSTL IO for Kintex-7.

13.4.1 Power analysis for SSTL 12

For the SSTL 12 IO, the SP of the device is 0.080 W which is 13% of the TP, and the DP is 0.518 W which is 81% of the TP. The TP is the sum of DP and SP, which is 0.598 W for SSTL 12. The TP consumption for SSTL 12 IO is represented in Figure 13.6.

We know that the SP is the summation of IO, S/G, and logic power. The IO, S/G, and logic power are 0.004, 0.417, and 0.097 W, respectively, for SSTL 12 IO. The graphical representation of on-chips power of the device is shown in Figure 13.7.

13.4.2 Power analysis for SSTL 135

For the SSTL 135 IO, the SP of the device is 0.100 W which is 1% of the TP, and the DP is 7.678 W which is the 99% of the TP. The TP is the sum of DP and SP, which is 7.778 W for SSTL 135. The TP consumption for SSTL 135 IO is represented in Figure 13.8.

On-Chip Power

	☐ Dynamic:	0.518 W (87%)
87% 81% 19%	▨ Signals:	0.417 W (81%)
	▨ Logic:	0.097 W (19%)
	☐ I/O:	0.004 W (0%)
13%	▨ Device Static:	0.080 W (13%)

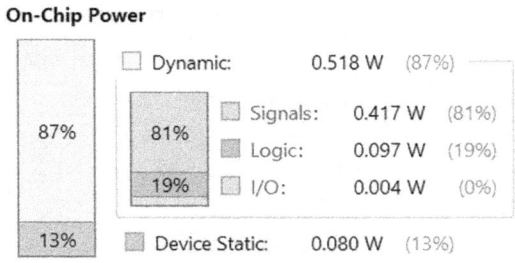

Figure 13.6 TP consumption for SSTL 12 IO.

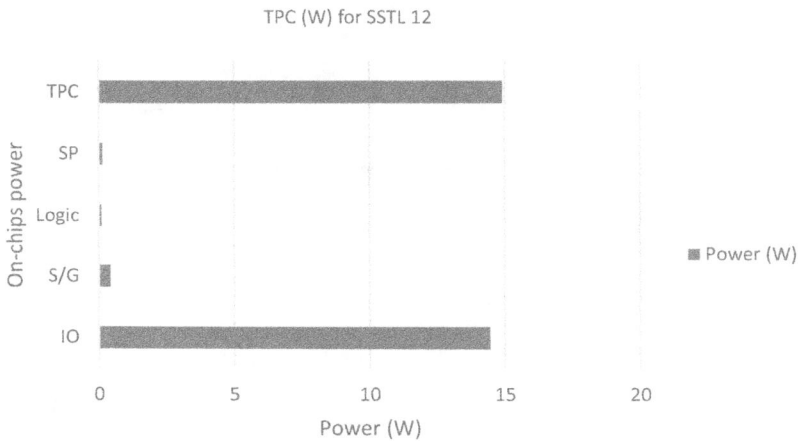

TPC (W) for SSTL 12

Figure 13.7 Graphical representation of on-chips power for SSTL 12 IO.

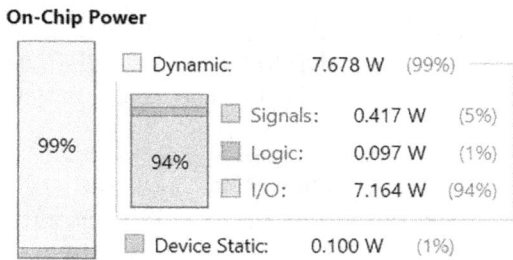

On-Chip Power

	☐ Dynamic:	7.678 W (99%)
99% 94%	▨ Signals:	0.417 W (5%)
	▨ Logic:	0.097 W (1%)
	☐ I/O:	7.164 W (94%)
	▨ Device Static:	0.100 W (1%)

Figure 13.8 TP consumption for SSTL 135 IO.

We know that the SP is the summation of IO, S/G, and logic power. The IO, S/G, and logic power are 7.164, 0.417, and 0.097 W, respectively, for SSTL 135 IO. The graphical representation of on-chips power of the device is shown in Figure 13.9.

TPC (W) for SSTL 135

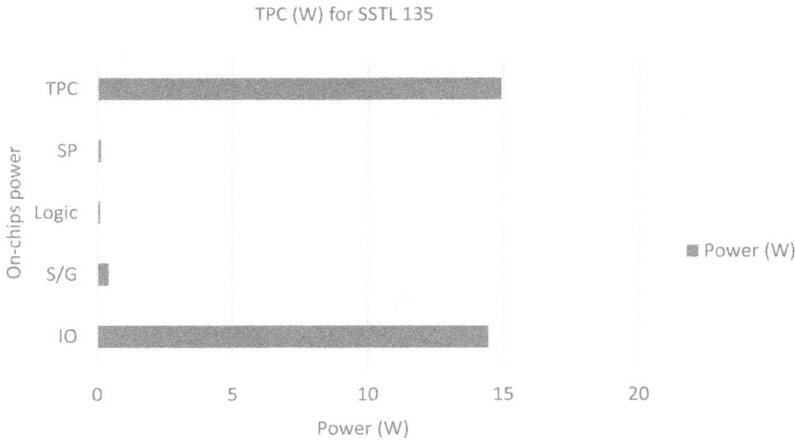

Figure 13.9 Graphical representation of on-chips power for SSTL 135 IO.

13.4.3 Power analysis for SSTL 15 IO

When the SSTL 15 IO standards are used for matching the impedance of the device, the TP consumption of the device observed is 7.956 W. The device SP is 0.102 W which is 1% of TP, and DP is 7.854 W which is 99% of TP consumption. The power consumption for SSTL15 IO is illustrated in Figure 13.10. The graphical representation of TP consumption is described in Figure 13.11.

13.4.4 Power analysis for SSTL 18_I

For the SSTL 18_I IO, the SP of the device is 0.124 W which is 1% of the TP, and the DP is 12.615 W which is the 99% of the TP. The TP is the sum of DP and SP, which is 12.739 W for SSTL 18_I. The TP consumption for SSTL 12 IO is represented in Figure 13.12.

We know that the SP is the summation of IO, S/G, and logic power. The IO, S/G, and logic power are 12.101, 0.417, and 0.097 W, respectively for SSTL 18_I IO. The leakage power is the DP which is 12.615 W. The graphical representation of on-chips power of the device is shown in Figure 13.13.

13.4.5 Power analysis for SSTL 18_II

For the SSTL 18_II IO, the SP of the device is 0.135 W which is 01% of the TP, and the DP is 14.767 W which is the 99% of the TP. The TP is the sum of DP and SP, which is 14.902 W for SSTL 18_II. The TP consumption for SSTL 18_II IO is represented in Figure 13.14.

On-Chip Power

☐ Dynamic:	7.854 W (99%)
■ Signals:	0.417 W (5%)
■ Logic:	0.097 W (1%)
☐ I/O:	7.340 W (94%)
■ Device Static:	0.102 W (1%)

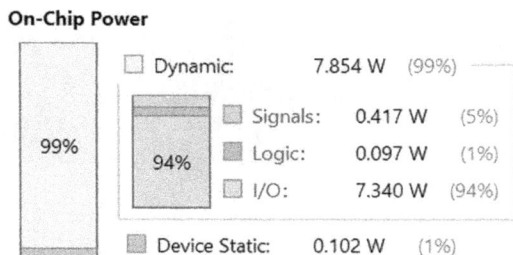

Figure 13.10 TP consumption for SSTL 15 IO.

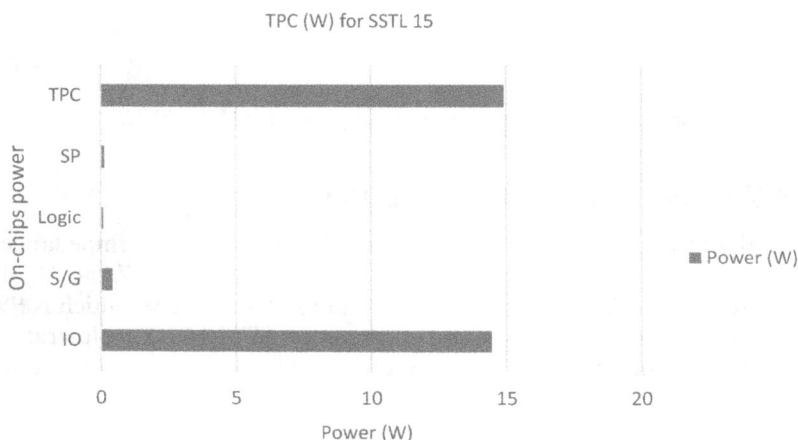

TPC (W) for SSTL 15

Figure 13.11 Graphical representation of on-chips power for SSTL 15 IO.

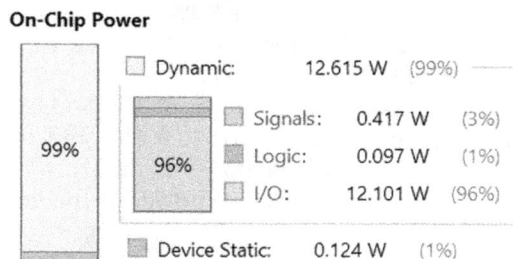

On-Chip Power

☐ Dynamic:	12.615 W (99%)
■ Signals:	0.417 W (3%)
■ Logic:	0.097 W (1%)
☐ I/O:	12.101 W (96%)
■ Device Static:	0.124 W (1%)

Figure 13.12 TP consumption for SSTL 18_I IO.

We know that the SP is the summation of IO, S/G, and logic power. The IO, S/G, and logic power are 14.253, 0.417, and 0.097 W, respectively, for SSTL 18_II IO. The leakage power is the DP which is 14.767 W. The graphical representation of on-chips power of the device is shown in Figure 13.15.

TPC (W) for SSTL 18_I

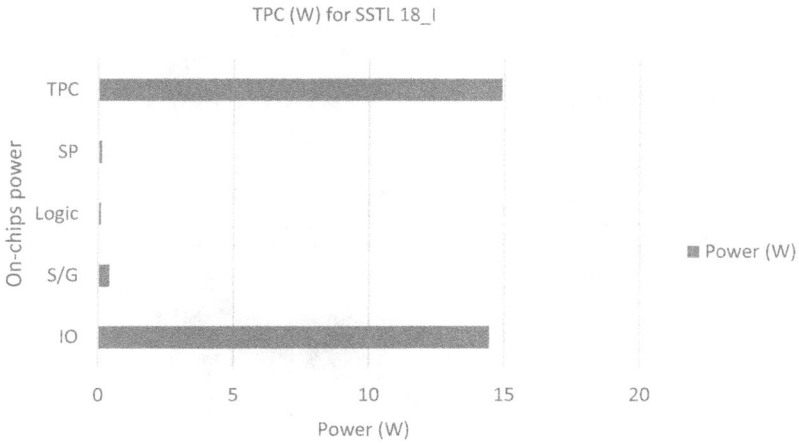

Figure 13.13 Graphical representation of on-chips power for SSTL 18_I IO.

On-Chip Power

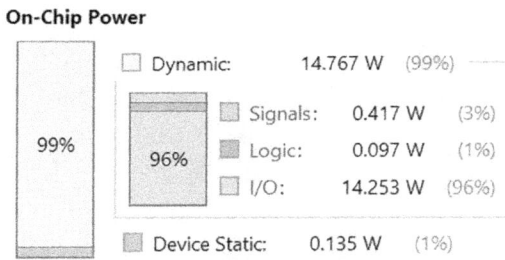

Figure 13.14 TP consumption for SSTL 18_II IO.

TPC (W) for SSTL 18_II

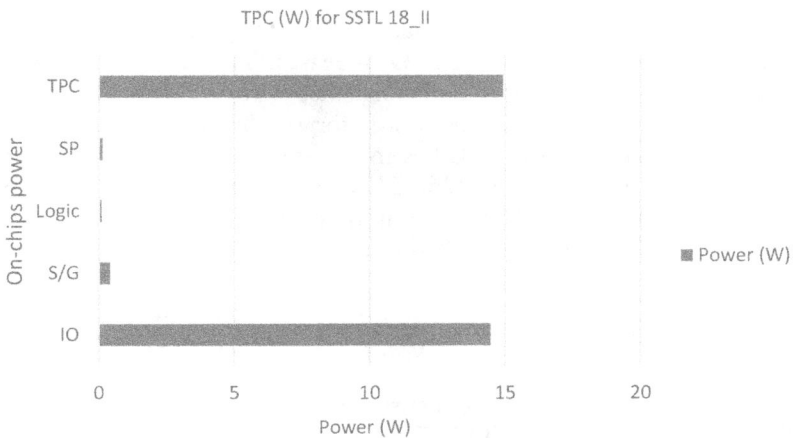

Figure 13.15 Graphical representation of on-chips power for SSTL 18_II IO.

Table 13.3 TPC for consumed IO standards

IO standards	TPC (W)
SSTL 12	0.598
SSTL 135	7.778
SSTL 15	7.956
SSTL 18_I	12.739
SSTL 18_II	14.902

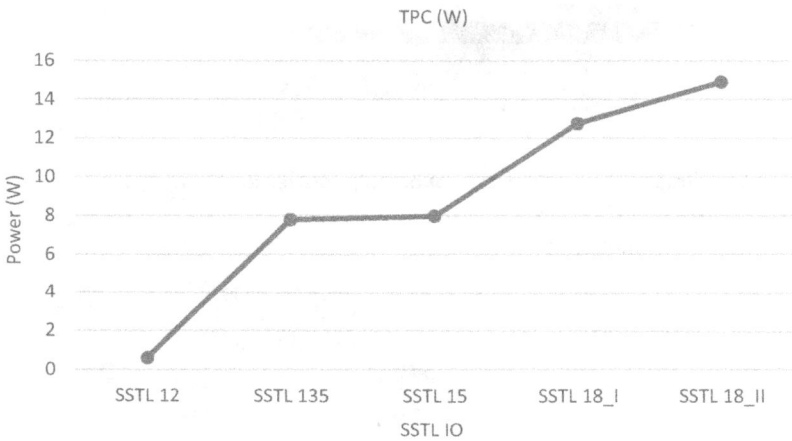

Figure 13.16 TP consumption for used IO standards.

13.5 OBSERVATION AND COMPARATIVE ANALYSIS

From Section 13.4, it is observed that the TPC of the device gets increased as the input voltage for the SSTL IO increases. The devices consume the optimal power when the impedance is matched with SSTL 12 IO (0.598 W). The device consumes maximum power for the SSTL 18_II IO (14.902). For the other SSTL IO, the TPC lies in between SSTL 12 and SSTL 18_II IO. There is an increment of 2256.52% in TPC for SSTL 12 and SSTL 18_II IO. The TPC for the SSTL IO family is depicted in Table 13.3, and the graphical representation is described in Figure 13.16.

13.6 CONCLUSION

This chapter presents a comprehensive examination of the TPC of the packet counter on the KINTEX-7 FPGA. The packet counter is one of the most important components used in signal processing, and it also has various DSP applications. This chapter analyses the PACKET COUNTER's power usage over KINTEX-7. PACKET COUNTER's design is applied on VIVADO ISE. Attempts are made to optimize the power consumption using SSTL IO. IO standards are used to match the impedance of the input line with the impedance of the output line. When impedance is perfectly matched, the power consumption of any device is optimized. Here, several SSTL IOs are used to match the impedance. As the input voltage of the IO standard increases, Section 13.4 demonstrates that the TPC likewise increases. Therefore, the device's power usage is optimized when the input voltage is low. The KINTEX-7 PACKET COUNTER design provides the low power usage when the impedance is matched with SSTL 12 IO. In terms of its future applicability, the PACKET COUNTER design may be implemented on a variety of additional FPGAs from notable families, including Artix-7, Spartan-7, Virtex-7, and many more. Additionally, the designs may be transformed to ASIC designs for improved applications and uses.

REFERENCES

1. Orlov, Sergei S., William Phillips, Eric Bjornson, Yuzuru Takashima, Padma Sundaram, Lambertus Hesselink, Robert Okas, Darren Kwan, and Raymond Snyder. "High-transfer-rate high-capacity holographic disk data-storage system." *Applied Optics* 43, no. 25 (2004): 4902–4914.
2. Rahmati, Ahmad, and Lin Zhong. "Context-for-wireless: context-sensitive energy-efficient wireless data transfer." In *Proceedings of the 5th International Conference on Mobile Systems, Applications and Services*, New York, pp. 165–178. ACM Press, 2007.
3. Oruganti, S. K., S. H. Heo, H. Ma, and Franklin Bien. "Wireless energy transfer-based transceiver systems for power and/or high-data rate transmission through thick metal walls using sheet-like waveguides." *Electronics Letters* 50, no. 12 (2014): 886–888.
4. Jambunathan, K., E. Lai, M.A. Moss, and B. L. Button. "A review of heat transfer data for single circular jet impingement." *International Journal of Heat and Fluid Flow* 13, no. 2 (1992): 106–115.
5. Klein, Randall W., Michael A. Temple, and Michael J. Mendenhall. "Application of wavelet-based RF fingerprinting to enhance wireless network security." *Journal of Communications and Networks* 11, no. 6 (2009): 544–555.
6. Sinha, Preeti, V. K. Jha, Amit Kumar Rai, and Bharat Bhushan. "Security vulnerabilities, attacks and countermeasures in wireless sensor networks at various layers of OSI reference model: A survey." In *2017 International Conference on Signal Processing and Communication (ICSPC)*, Limassol, Cyprus, pp. 288–293. IEEE, 2017.
7. Banerjee, Suman, Bobby Bhattacharjee, and Christopher Kommareddy. "Scalable application layer multicast." In *Proceedings of the 2002 Conference on Applications, Technologies, Architectures, and Protocols for Computer Communications*, Pittsburg, PA, pp. 205–217. ACM Press, 2002.

8. Gupta, Isha, Swati Singh Garima, Harpreet Kaur, Deepshikha Bhatt, and Aamir Vohra. "28nm FPGA based power optimized UART design using HSTL I/O standards." *Indian Journal of Science and Technology* 8, no. 17 (2015): 1–6.

9. Kumar, Keshav, Amanpreet Kaur, S. N. Panda, and Bishwajeet Pandey. "Effect of different nano meter technology-based FPGA on energy efficient UART design." In *2018 8th International Conference on Communication Systems and Network Technologies (CSNT)*, India, pp. 1–4. IEEE, 2018.

10. Kumar, Keshav, Amanpreet Kaur, Bishwajeet Pandey, and S. N. Panda. "Low power UART design using different nanometer technology based FPGA." In *2018 8th International Conference on Communication Systems and Network Technologies (CSNT)*, India, pp. 1–3. IEEE, 2018.

11. Kumar, Keshav, Bishwajeet Pandey, Amit Kant Pandit, Y. A. Baker El-Ebiary, Salameh A. Mjlae, and Samer Bamansoor. "Design of low power transceiver on Spartan-3 and Spartan-6 FPGA." *International Journal of Innovative Technology and Exploring Engineering* 8, no. 12S2 (2019): 27–30.

12. Sandhu, Amanpreet, Vidhoytma Gandhi, Simranpreet Kaur, Surbhi Huria, Divjot Singh, and Wamika Goyal. "Thermally aware LVCMOS based low power universal asynchronous receiver transmitter design on FPGA." *Indian Journal of Science and Technology* 8, no. 20 (2015): 1–4.

13. Kumar, Abhishek, Bishwajeet Pandey, D. M. Akbar Hussain, Mohammad Atiqur Rahman, Vishal Jain, and Ayoub Bahanasse. "Low voltage complementary metal oxide semiconductor-based energy efficient UART design on Spartan-6 FPGA." In *2019 11th International Conference on Computational Intelligence and Communication Networks (CICN)*, India, pp. 84–87. IEEE, 2019.

14. Pandey, Bishwajeet, and Ravikant Kumar. "Low voltage DCI based low power VLSI circuit implementation on FPGA." In *2013 IEEE Conference on Information & Communication Technologies*, India, pp. 128–131. IEEE, 2013.

15. Kaur, Harkinder, Harsh Sohal, and Jaiteg Singh. "Design and performance analysis of uart using Altera Quartus-II and Xilinx ISE 14.2." In *6th International Conference on Communication and Network Technologies*, India. 2016.

16. Kaur, Ravinder, Jagdish Kumar, Sumita Nagah, Bishwajeet Pandey, and Kavita Goswami. "IO Standard based low power memory design and implementation on FPGA." In *2015 2nd International Conference on Computing for Sustainable Global Development (INDIACom)*, India, pp. 1501–1505. IEEE, 2015.

17. Kumar, Vivek, Aksh Rastogi, and V. K. Tomar. "Implementation of UART Design for RF Modules Using Different FPGA Technologies." *IOP Conference Series: Materials Science and Engineering* 1116, no. 1 (2021): 012131.

Chapter 14

HSTL-based packet
counter for GCC

LIST OF ABBREVIATIONS

AC	Alternating Current
ASIC	Application-Specific Integrated Circuit
BUFG	Global Buffers
Clk	Clocks
CMOS	Complementary Metal-Oxide-Semiconductor
DDR	Double Data Rate
DRAM	Dynamic Random-Access Memory
DP	Dynamic Power
EB	Exabytes
FF	Flip-Flops
FPGA	Field Programmable Gate Array
GC	Green Computing
GCC	Green Communication Computing
G. Comm.	Green Communication
GPS	Global Positioning System
HSTL	High-Speed Transceiver Logic
IC	Integrated Circuit
IO	Input Output
IT	Information Technology
JT	Junction Temperature
LP	Leakage Power
LUT	Look Up Tables
LVCMOS	Low-Voltage Complementary Metal-Oxide-Semiconductor
RTL	Register Transfer Logic
S/G	Signal
SP	Static Power
TCP/IP	Transmission Control Protocol/Internet Protocol
TM	Thermal Margin
TP	Total Power
TPC	Total Power Consumption
θJA	Effective Thermal Resistance to Air

DOI: 10.1201/9781003302872-14

14.1 INTRODUCTION

In the era of information technology (IT), it seems to be relatively simple to effectively transport data from one node to another node or one device to another device. The data might be sent in a matter of seconds. The transport of data from one device to another is accomplished by packets. These packets are often referred to as data packets or just packets. Packet transport is only possible with the use of networking principles. When data is transported via the Internet, there is a possibility that some packets may be lost or arrive at their destination late [1,2]. Consequently, in the era of information technology, the absence or delay of such data may result in serious concerns for certain organizations. Using the "packet counter," it is possible to resolve certain types of data transmission issues. The packet counter is software or a program that, when loaded in advance on the PC doing data communications, determines in real time the data transmission size and estimated data communications expenses of PacketWIN [3,4]. It monitors the transit of data packets from the source to the destination. Transferring the data packet from its source to its destination is referred to as computer networking. The transmission of data packets is performed via the various TCP/IP levels. As seen in Figure 14.1, the TCP/IP paradigm consists of five levels on average.

The packets counter monitors and counts every packet passed from the physical layer to the application layer. Communication happens at the application layer. The packet counter supports the transmission of data packets between the physical and application layers [5,6]. In the hardware implementation of the packet counter, the device may use a substantial amount of energy. Massive power use necessitates massive resource usage, thus affecting the device's durability. To guarantee the minimal use of natural resources and to improve the device's life cycle, one should implement GC principles. Additionally, GC contributes to sustainable development.

This chapter focuses on the development of a power-efficient packet counter paradigm that not only supports efficient data packet transfer but also guarantees GC. A power-efficient sort of packet counter is constructed

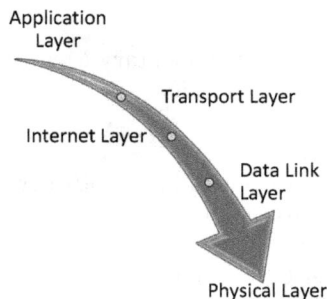

Figure 14.1 TCP/IP layer.

using the FPGA device. FPGAs are semiconductor components used to construct digital circuits. The FPGA device may be customized after it has been manufactured, which is its major advantage [7–9]. We are utilizing Kintex-7 FPGA with HSTL IO to match the impedance of the circuitry in order to make it appropriate for GC. This chapter explores how the FPGA device may be utilized to make the packet counter energy-efficient.

14.2 IMPLEMENTATION OF PACKET COUNTER ON FPGA

For the development of a power-efficient packet counter model, the Kintex-7 FPGA device was utilized. It is a Xilinx family device with a 28 nm gate size. Using the VIVADO ISE, the packet counter design is implemented on the same device. To optimize power consumption, HSTL_I IOs are employed. IO standard matches the transmission line impedance for the internal circuit in order to minimize power consumption. To implement the design on the FPGA device, FPGA resources such as IO, FF, LUTs, and BUFG are utilized [10–12]. The resource utilization for packet counter design is displayed in Table 14.1 and illustrated in Figure 14.2.

Table 14.1 Resource utilization for designing packet counter design

Resources	Available	Utilization	Utilization (%)
LUT	4	41,000	0.01
IO	71	300	23.67
BUFG	1	32	3.13
FF	38	82,000	0.05

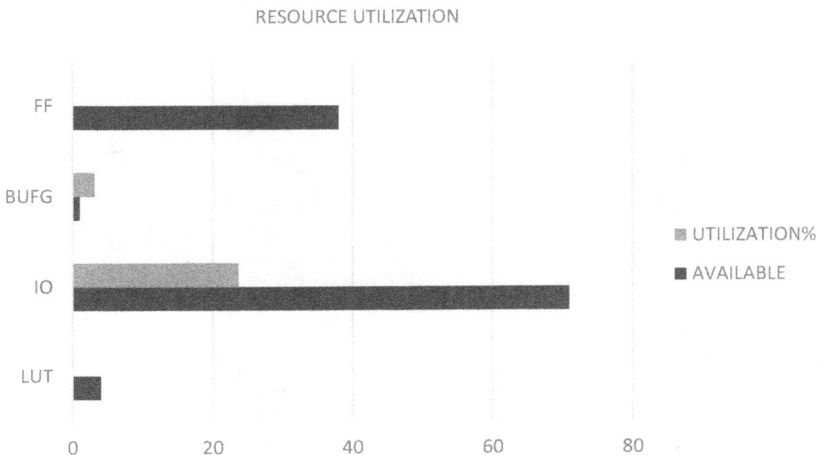

Figure 14.2 Resource utilization for designing packet counter design.

From Table 14.1, it is observed that in the process of designing the packet counter design, 71 IOs and 4 LUTs are consumed as well as 38 FFs and 1 BUFG of Kintex-7 device. The RTL design of packet counter is shown in Figure 14.3, while the technology schematic of the packet counter is described in Figure 14.4.

Figure 14.3 describes the RTL of the packet counter. In Figure 14.3, it is observed that there are two blocks of counter which counts the bits. At the input end, we have a clk pulse generator, along with a 32-bit packet input. At the output end, the counter counts the bits according to the clk pulse and gives the packet outputs (32-bits).

14.3 THERMAL PROPERTIES

This section will give a brief understanding about the thermal properties of the device. The thermal properties are taken into account while matching the impedance using the LVCMOS IO standard [13]. The thermal properties considered here are as follows:

a. TM: This is a device's attribute that permits it to consume very little power.
b. ϑJA: When 1 W of power is dissipated in the IC, the amount of heat created or a rise in temperature is specified.
c. JT: The highest operating temperature of an FPGA device's integrated gates.

The thermal properties for the Kintex-7 device, when the impedance is matched with LVCMOS IO, are well described in Table 14.2 and Figure 14.5.

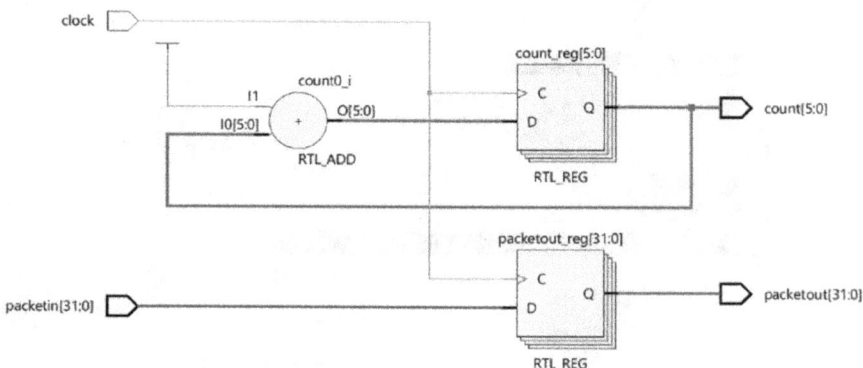

Figure 14.3 RTL schematic of packet counter.

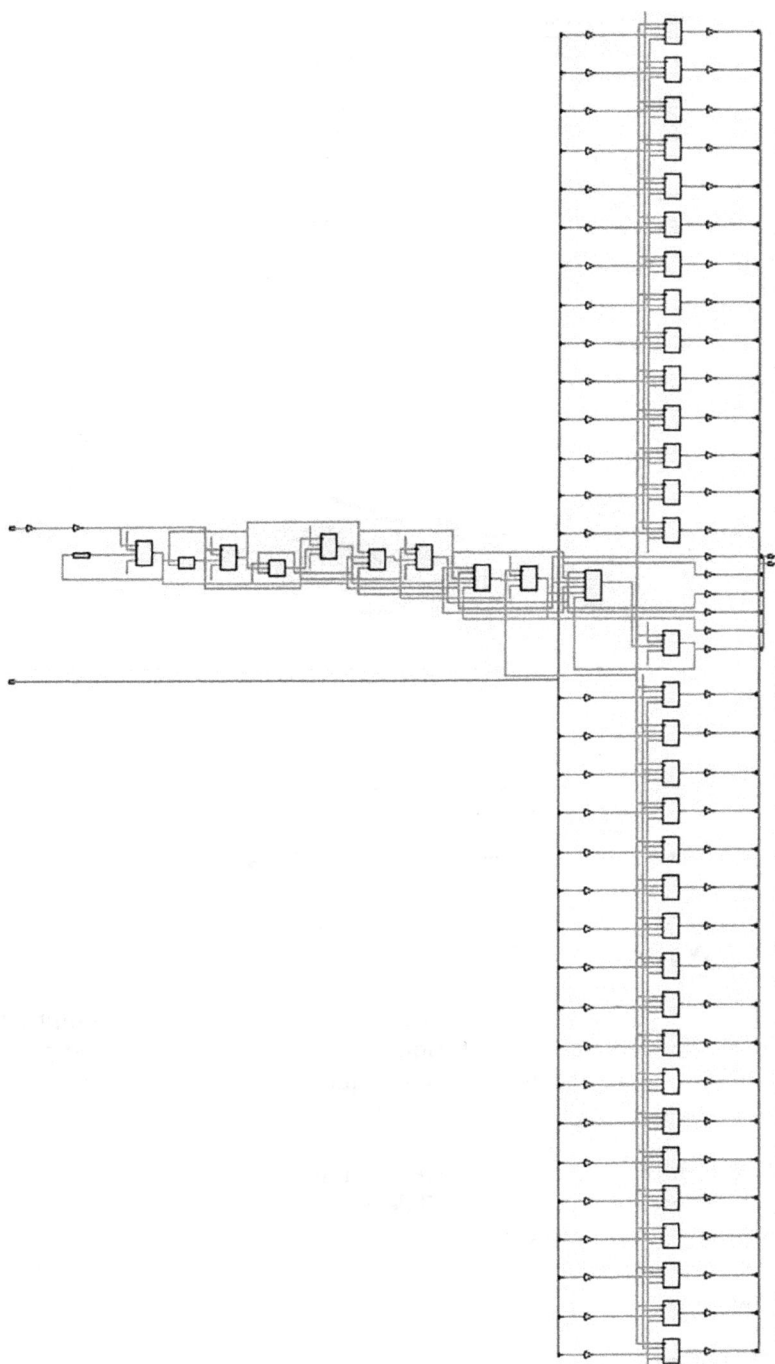

Figure 14.4 Technology schematic of packet counter.

Table 14.2 Thermal properties for the HSTL IO for Kintex-7

IO std	JT (°C)	TM (°C)	ϑJA (°C/W)
HSTL_I	46.2	53.8	1.9
HSTL_II	37.7	62.3	1.9
HSTL_I_18	50.8	49.2	1.9
HSTL_II_18	38.7	61.3	1.9
HSTL_I_12	42.1	57.9	1.9

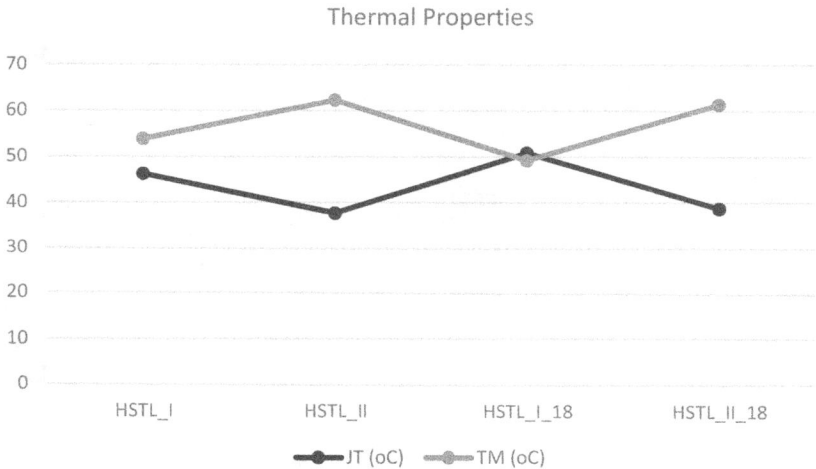

Figure 14.5 Thermal properties for the HSTL IO for Kintex-7.

14.4 POWER ANALYSIS

The world's greatest threat is becoming the energy consumption of electronic gadgets and information technology. Currently, the globe is suffering an energy scarcity. In consequence, the whole ecosystem collaborates to develop a low-power solution for the information technology and electronics sectors. Reduced power consumption increases the device's durability. In this part, we calculate the power consumption of a FIR implementation on a Kintex-7 device using the HSTL_I IO Standard. The device's total power consumption is the sum of both dynamic and static power consumption [14–17]. This is described mathematically by Equation 14.1:

$$TP = SP + DP \tag{14.1}$$

where

TP = Total Power
SP = Static Power
DP = Dynamic Power

14.4.1 Power analysis for HSTL_I

For the HSTL_I IO, the SP of the device is 0.115 W which is 1% of the TP, and the DP is 11.144 W which is 99% of the TP. The TP is the sum of DP and SP, which is 11.259 W for HSTL_I. The TP consumption for HSTL_I IO is represented in Figure 14.6.

We know that the SP is the summation of IO, S/G, and logic power. The IO, S/G, and logic power are 10.630, 0.417, and 0.097 W, respectively, for HSTL_I IO. The graphical representation of on-chips power of the device is shown in Figure 14.7.

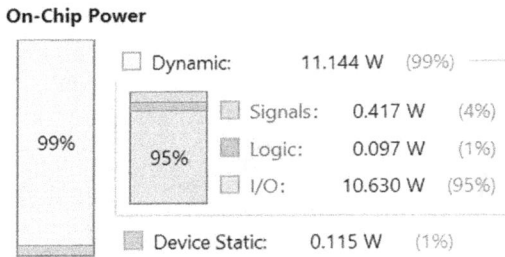

On-Chip Power

□ Dynamic:	11.144 W	(99%)	
▨ Signals:	0.417 W	(4%)	
▨ Logic:	0.097 W	(1%)	
□ I/O:	10.630 W	(95%)	
▨ Device Static:	0.115 W	(1%)	

(99% / 95%)

Figure 14.6 TP consumption for HSTL_I IO.

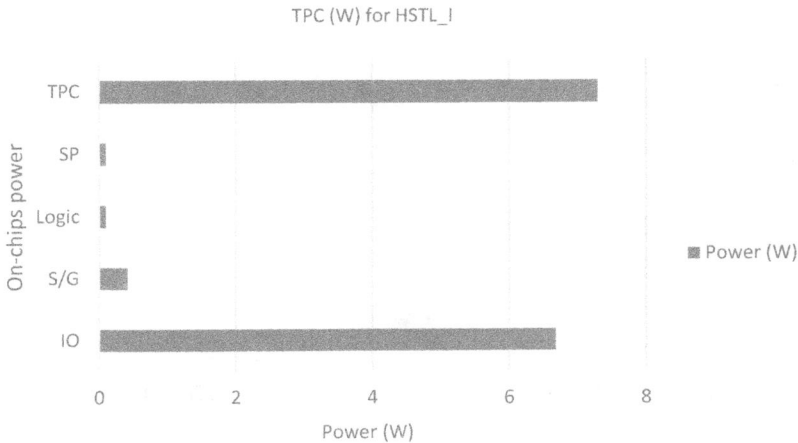

TPC (W) for HSTL_I

On-chips power: TPC, SP, Logic, S/G, IO

■ Power (W)

Power (W)

Figure 14.7 Graphical representation of on-chips power for HSTL_I IO.

14.4.2 Power analysis for HSTL_II

For the HSTL_II IO, the SP of the device is 0.099 W which is 1% of the TP, and the DP is 6.639 W which is 99% of the TP. The TP is the sum of DP and SP, which is 6.738 W for HSTL_II. The TP consumption for HSTL_II IO is represented in Figure 14.8.

We know that the SP is the summation of IO, S/G, and logic power. The IO, S/G, and logic power are 6.125, 0.417, and 0.097 W, respectively, for HSTL_II IO. The graphical representation of on-chips power of the device is shown in Figure 14.9.

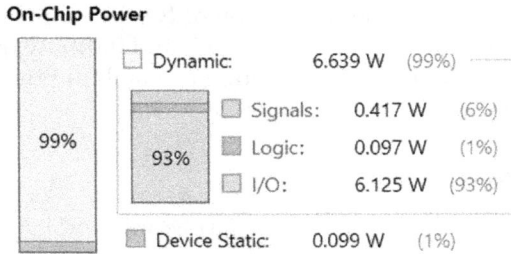

Figure 14.8 TP consumption for HSTL_II IO.

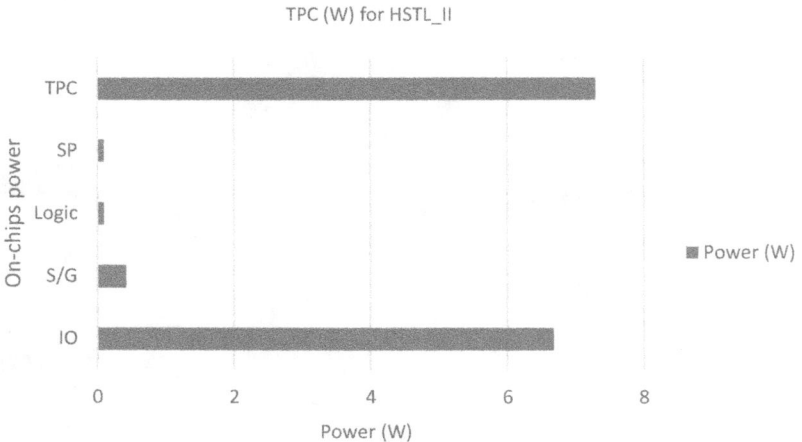

Figure 14.9 Graphical representation of on-chips power for HSTL_II IO.

14.4.3 Power analysis for HSTL_I_12 IO

When the HSTL_I_12 IO standard is used for matching the impedance of the device, the TP consumption of the device observed is 9.066 W. The device SP is 0.108 W which is 1% of TP, and DP is 8.958 W which is 99% of TP consumption. The power consumption for HSTL_I_12 IO is illustrated in Figure 14.10. The graphical representation of TP consumption is described in Figure 14.11.

14.4.4 Power analysis for HSTL_I_18

For the HSTL_I_18 IO, the SP of the device is 0.128 W which is 1% of the TP, and the DP is 13.547 W which is 99% of the TP. The TP is the sum of DP and SP, which is 13.675 W for HSTL_I_18. The TP consumption for HSTL 12 IO is represented in Figure 14.12.

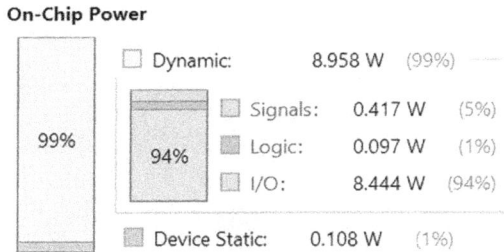

On-Chip Power

□ Dynamic:	8.958 W	(99%)
▨ Signals:	0.417 W	(5%)
▨ Logic:	0.097 W	(1%)
▨ I/O:	8.444 W	(94%)
▨ Device Static:	0.108 W	(1%)

99% 94%

Figure 14.10 TP consumption for HSTL_I_12 IO.

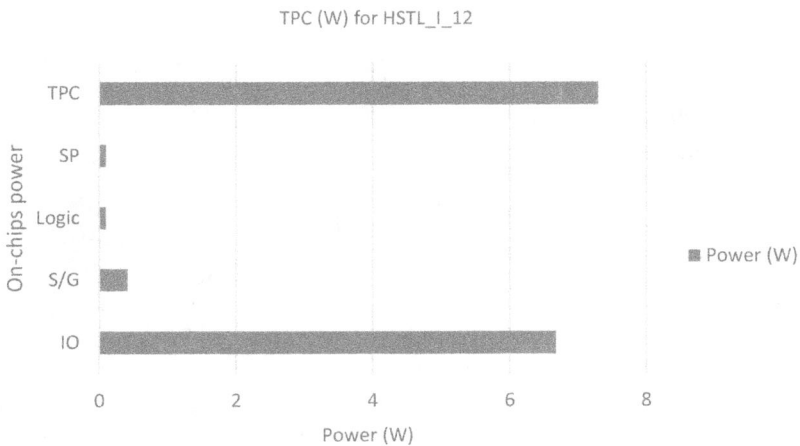

TPC (W) for HSTL_I_12

On-chips power: TPC, SP, Logic, S/G, IO

■ Power (W)

Power (W)

Figure 14.11 Graphical representation of on-chips power for HSTL_I_12 IO.

On-Chip Power

☐ Dynamic: 13.547 W   (99%)

▨ Signals: 0.417 W   (3%)

▨ Logic: 0.097 W   (1%)

☐ I/O: 13.033 W   (96%)

99%   96%

▨ Device Static: 0.128 W   (1%)

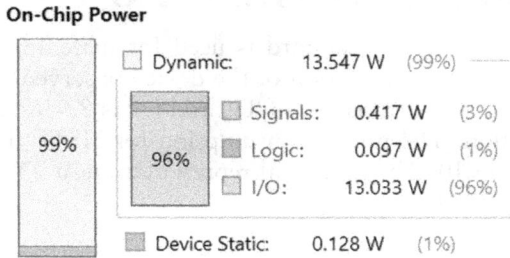

Figure 14.12 TP consumption for HSTL_I_18 IO.

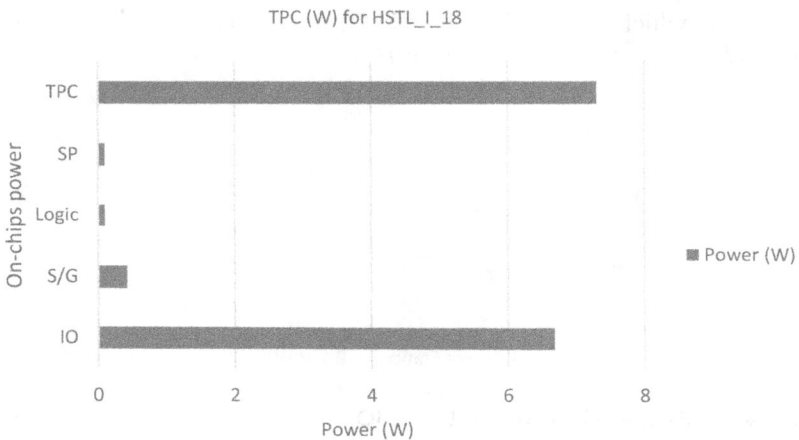

TPC (W) for HSTL_I_18

Figure 14.13 Graphical representation of on-chips power for HSTL_I_18 IO.

We know that the SP is the summation of IO, S/G, and logic power. The IO, S/G, and logic power are 13.033, 0.417, and 0.097 W, respectively, for HSTL_I_18 IO. The leakage power is the DP which is 13.547 W. The graphical representation of on-chips power of the device is shown in Figure 14.13.

14.4.5 Power analysis for HSTL_II_18

For the HSTL_II_18 IO, the SP of the device is 0.100 W which is 1% of the TP, and the DP is 7.193 W which is 99% of the TP. The TP is the sum of DP and SP, which is 7.293 W for HSTL_II_18. The TP consumption for HSTL_II_18 is represented in Figure 14.14.

We know that the SP is the summation of IO, S/G, and logic power. The IO, S/G, and logic power are 6.679, 0.417, and 0.097 W, respectively, for HSTL_II_18 IO. The leakage power is the DP which is 7.193 W. The graphical representation of on-chips power of the device is shown in Figure 14.15.

On-Chip Power

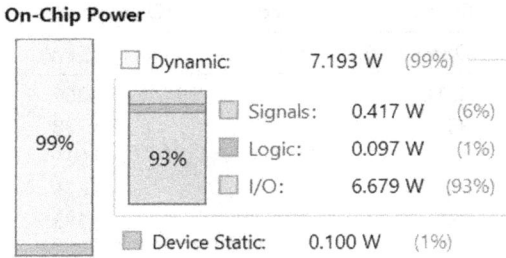

☐ Dynamic:	7.193 W (99%)
☐ Signals:	0.417 W (6%)
☐ Logic:	0.097 W (1%)
☐ I/O:	6.679 W (93%)
☐ Device Static:	0.100 W (1%)

Figure 14.14 TP consumption for HSTL_II_18 IO.

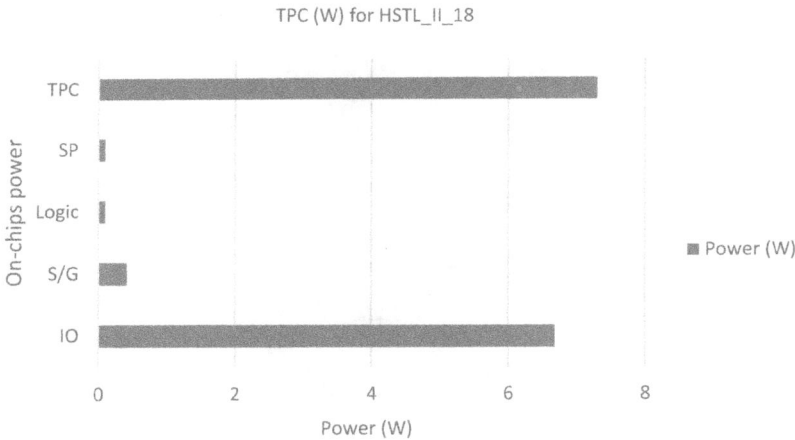

Figure 14.15 Graphical representation of on-chips power for HSTL_II_18 IO.

14.5 OBSERVATION AND COMPARATIVE ANALYSIS

From Section 14.4, it is observed that the TPC of the device gets changed for every impedance matching with the HSTL IO. The devices consume the optimal power when the impedance is matched with HSTL_II IO (6.738 W). The device consumes maximum power for the HSTL_I_18 IO (13.675). For the other HSTL IO, the TPC lies in between HSTL_II and HSTL_I_18 IO. There is an increment of 102.953% in TPC for HSTL_II and HSTL_I_18 IO. The TPC for the HSTL IO family is depicted in Table 14.3, and the graphical representation is described in Figure 14.16.

Table 14.3 TPC for consumed IO standards

IO standards	TPC (W)
HSTL_I	11.259
HSTL_II	6.738
HSTL_I_12	9.066
HSTL_I_18	13.675
HSTL_II_18	7.293

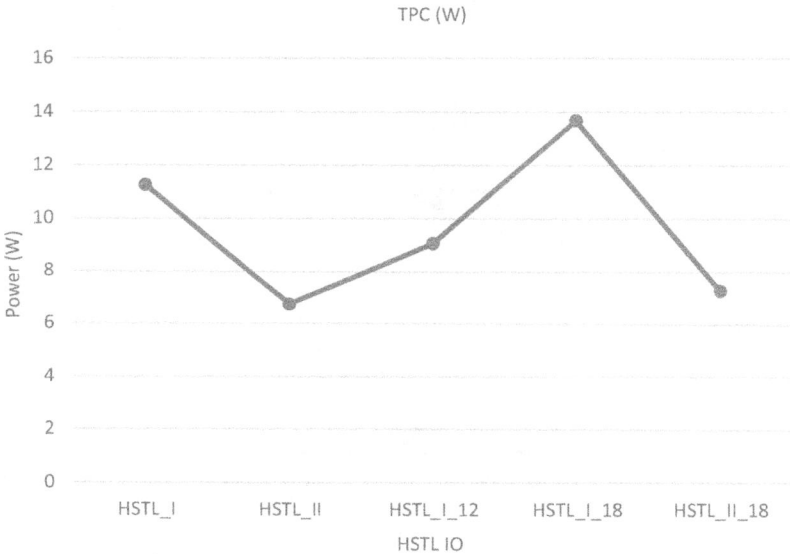

Figure 14.16 TP consumption for used IO standards.

14.6 CONCLUSION

This chapter provides an in-depth analysis of the TPC of the packet counter on the KINTEX-7 FPGA. The packet counter is one of the most significant signal processing components, and it has several DSP applications. This chapter examines the power consumption of the packet counter over KINTEX-7. The design of packet counter is applied to VIVADO ISE. Utilizing HSTL IO, attempts are made to optimize the power usage. IO standards are used to match the input line impedance to the output line impedance. When impedance is correctly matched, any device's power usage is optimized. Multiple HSTL IOs are used here to match the impedance. As the impedance matching with HSTL IO varies, the TPC of the devices corresponding to each HSTL IO also varies. The KINTEX-7 packet counter design offers low power consumption when coupled with HSTL II IO impedance. Regarding its future application, the packet counter architecture may be implemented on a number of additional FPGAs from significant families, such as Artix-7, Spartan-7, and Virtex-7. In addition, the designs may be converted into ASIC designs for enhanced applications and purposes.

REFERENCES

1. Whiting, Doug, Russ Housley, and Niels Ferguson. Counter with CBC-MAC (CCM). No. rfc3610. 2003.
2. Housley, Russell. Using advanced encryption standard (AES) counter mode with IPsec encapsulating security payload (ESP). No. rfc3686. 2004.
3. Packet counter | Software download. au. https://www.au.com/english/mobile/service/mobile-communications/soft-download/packet-counter/. Accessed on 30 June 2023.
4. Yang, Jianwei, Fan Dai, Jielin Wang, Jianmin Zeng, Zhang Zhang, Jun Han, and Xiaoyang Zeng. "Countering power analysis attacks by exploiting characteristics of multicore processors." *IEICE Electronics Express* 15, no. 7 (2018). https://doi.org/10.1587/elex.15.20180084.
5. Kaur, Damandeep, and Parminder Singh. "Various OSI layer attacks and countermeasure to enhance the performance of WSNs during wormhole attack." *International Journal on Network Security* 5, no. 1 (2014): 62.
6. Klein, Randall W., Michael A. Temple, and Michael J. Mendenhall. "Application of wavelet-based RF fingerprinting to enhance wireless network security." *Journal of Communications and Networks* 11, no. 6 (2009): 544–555.
7. Sinha, Preeti, V. K. Jha, Amit Kumar Rai, and Bharat Bhushan. "Security vulnerabilities, attacks and countermeasures in wireless sensor networks at various layers of OSI reference model: A survey." In *2017 International Conference on Signal Processing and Communication (ICSPC)*, Limassol, Cyprus, pp. 288–293. IEEE, 2017.
8. Banerjee, Suman, Bobby Bhattacharjee, and Christopher Kommareddy. "Scalable application layer multicast." In *Proceedings of the 2002 Conference on Applications, Technologies, Architectures, and Protocols for Computer Communications*, Pittsburg, PA, pp. 205–217. ACM Press, 2002.
9. Gupta, Isha, Swati Singh Garima, Harpreet Kaur, Deepshikha Bhatt, and Aamir Vohra. "28nm FPGA based power optimized UART design using HSTL I/O standards." *Indian Journal of Science and Technology* 8, no. 17 (2015): 1–6.
10. Kumar, Keshav, Amanpreet Kaur, S. N. Panda, and Bishwajeet Pandey. "Effect of different nano meter technology-based FPGA on energy efficient UART design." In *2018 8th International Conference on Communication Systems and Network Technologies (CSNT)*, India, pp. 1–4. IEEE, 2018.
11. Kumar, Keshav, Amanpreet Kaur, Bishwajeet Pandey, and S. N. Panda. "Low power UART design using different nanometer technology-based FPGA." In *2018 8th International Conference on Communication Systems and Network Technologies (CSNT)*, India, pp. 1–3. IEEE, 2018.

12. Kumar, Keshav, Bishwajeet Pandey, Amit Kant Pandit, Y. A. Baker El-Ebiary, Salameh A. Mjlae, and Samer Bamansoor. "Design of low power transceiver on Spartan-3 and Spartan-6 FPGA." *International Journal of Innovative Technology and Exploring Engineering* 8, no. 12S2 (2019): 27–30.

13. Sandhu, Amanpreet, Vidhoytma Gandhi, Simranpreet Kaur, Surbhi Huria, Divjot Singh, and Wamika Goyal. "Thermally aware LVCMOS based low power universal asynchronous receiver transmitter design on FPGA." *Indian Journal of Science and Technology* 8, no. 20 (2015): 1–4.

14. Kumar, Abhishek, Bishwajeet Pandey, D. M. Akbar Hussain, Mohammad Atiqur Rahman, Vishal Jain, and Ayoub Bahanasse. "Low voltage complementary metal oxide semiconductor-based energy efficient UART design on Spartan-6 FPGA." In *2019 11th International Conference on Computational Intelligence and Communication Networks (CICN)*, India, pp. 84–87. IEEE, 2019.

15. Pandey, Bishwajeet, and Ravikant Kumar. "Low voltage DCI based low power VLSI circuit implementation on FPGA." In *2013 IEEE Conference on Information & Communication Technologies*, India, pp. 128–131. IEEE, 2013.

16. Kaur, Harkinder, Harsh Sohal, and Jaiteg Singh. "Design and performance analysis of uart using Altera Quartus-II and Xilinx ISE 14.2." In *6th International Conference on Communication and Network Technologies*, India. 2016.

17. Kaur, Ravinder, Jagdish Kumar, Sumita Nagah, Bishwajeet Pandey, and Kavita Goswami. "IO Standard based low power memory design and implementation on FPGA." In *2015 2nd International Conference on Computing for Sustainable Global Development (INDIACom)*, India, pp. 1501–1505. IEEE, 2015.

Chapter 15

MOBILE DDR-based packet counter for GCC

LIST OF ABBREVIATIONS

AC	Alternating Current
ASIC	Application-Specific Integrated Circuit
BUFG	Global Buffers
Clk	Clocks
CMOS	Complementary Metal-Oxide-Semiconductor
DDR	Double Data Rate
DRAM	Dynamic Random-Access Memory
DP	Dynamic Power
EB	Exabytes
FF	Flip-Flops
FPGA	Field Programmable Gate Array
GC	Green Computing
GCC	Green Communication Computing
G. Comm.	Green Communication
GPS	Global Positioning System
IC	Integrated Circuit
IO	Input Output
IT	Information Technology
JT	Junction Temperature
LP	Leakage Power
LUT	Look Up Tables
LVCMOS	Low-Voltage Complementary Metal-Oxide-Semiconductor
RTL	Register Transfer Logic
S/G	Signal
SP	Static Power
TCP/IP	Transmission Control Protocol/Internet Protocol
TM	Thermal Margin
TP	Total Power
TPC	Total Power Consumption
θJA	Effective Thermal Resistance to Air

DOI: 10.1201/9781003302872-15

15.1 INTRODUCTION

In the era of information technology (IT), it seems to be relatively simple to effectively transport data from one node to another node or one device to another device. The data might be sent in a matter of seconds. The transport of data from one device to another is accomplished by packets. These packets are often referred to as data packets or just packets. Packet transport is only possible with the use of networking principles [1]. When data is transported via the Internet, there is a possibility that some packets may be lost or arrive at their destination late. Consequently, in the era of information technology, the absence or delay of such data may result in serious concerns for certain organizations [2].

Similar to Moore's law, it seems that the amount of digital data stored is basically increasing exponentially. According to Kryder's law, the amount of available storage capacity seems to be expanding practically exponentially. In 1986, the technological capacity of the world to receive information via one-way broadcast networks was 432 exabytes of information (optimally compressed); in 1993, it was 715 exabytes; in 2000, it was 1.2 zettabytes; and in 2007, it was 1.9 zettabytes, the information equivalent of 174 newspapers per person per day [3,4].

Such kinds of problems occurring in the data transmission can be rectified using the "packet counter." In the world of Internet, the data packets transferred from sender to receiver use the TCP/IP protocols. There are basically five different layers in the TCP/IP model which is described in Figure 15.1.

The packets counter monitors and counts every packet passed from the physical layer to the application layer. Communication happens at the application layer. The packet counter supports the transmission of data packets between the physical and application layers [5–7]. In the hardware implementation of the packet counter, the device may use a substantial amount of energy. Massive power use necessitates massive resource usage, hence affecting the device's durability. To guarantee the minimal use of natural

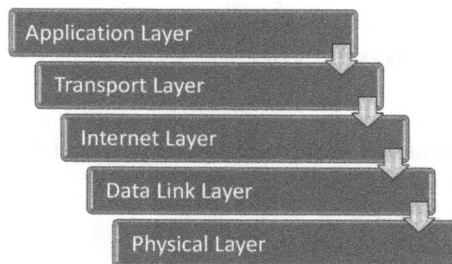

Figure 15.1 TCP/IP layer.

resources and to improve the device's life cycle, one should implement GC principles. Additionally, GC contributes to sustainable development [8].

This chapter focuses on the development of a power-efficient packet counter paradigm that not only supports efficient data packet transfer but also guarantees GC. A power-efficient sort of packet counter is constructed using the FPGA device. FPGAs are semiconductor components used to construct digital circuits. The FPGA device may be customized after it has been manufactured, which is its major advantage [9,10]. We are utilizing Kintex-7 FPGA with MOBILE DDR IO to match the impedance of the circuitry in order to make it appropriate for GC. In this chapter, we will examine how the FPGA device may be utilized to reduce the packet counter's power consumption for different frequency values with MOBILE DDR IO.

15.2 IMPLEMENTATION OF PACKET COUNTER ON FPGA

Utilizing the Kintex-7 FPGA chip, a power-efficient packet counter model was created. It is a device from the Xilinx family with a 28 nm gate size. The packet counter design is implemented on the same device using the VIVADO ISE. MOBILE DDR IOs are used to optimize power usage. In order to minimize power consumption, the IO standard matches the transmission line impedance to the internal circuit. FPGA resources including IO, FF, LUTs, and BUFG are used to realize the design on the FPGA device [11–13]. The use of resources for the packet counter design is given in Table 15.1 and depicted in Figure 15.2.

In creating the packet counter, 71 IOs, 4 LUTs, 38 FFs, and 1 BUFG of the Kintex-7 device are used, as shown in Table 15.1. Figure 15.3 depicts the RTL design of the packet counter, whereas Figure 15.4 depicts the technological schematic of the packet counter.

Figure 15.3 describes the RTL of the packet counter. In Figure 15.3, it is observed that there are two blocks of counter which counts the bits. At the input end, we have a clk pulse generator, along with a 32-bit packet input. At the output end, the counter counts the bits according to the clk pulse and gives the packet outputs (32-bits).

Table 15.1 Resource utilization for designing packet counter design

Resources	Available	Utilization	Utilization (%)
LUT	4	41,000	0.01
IO	71	300	23.67
BUFG	1	32	3.13
FF	38	82,000	0.05

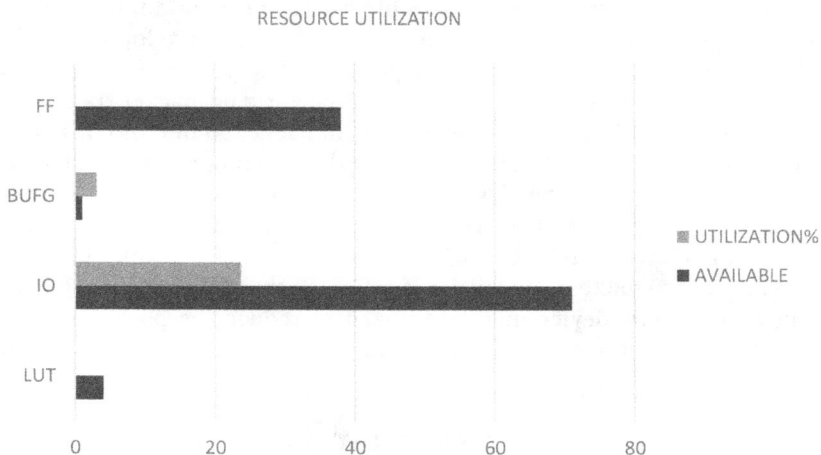

Figure 15.2 Resource utilization for designing packet counter design.

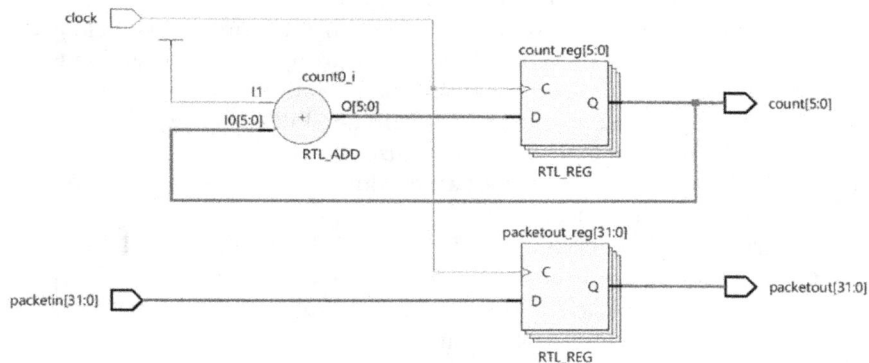

Figure 15.3 RTL schematic of packet counter.

15.3 THERMAL PROPERTIES

This section will provide an overview of the thermal characteristics of the device. The thermal properties are taken into account while matching impedance using the MOBILE DDR IO standard. The following thermal characteristics are taken into account:

a. TM: This is a feature of a device that enables very low power consumption.
b. ϑJA: When 1 W of power is dissipated in the IC, the amount of heat created or a rise in temperature is specified.
c. JT: The highest operating temperature of an FPGA device's integrated gates.

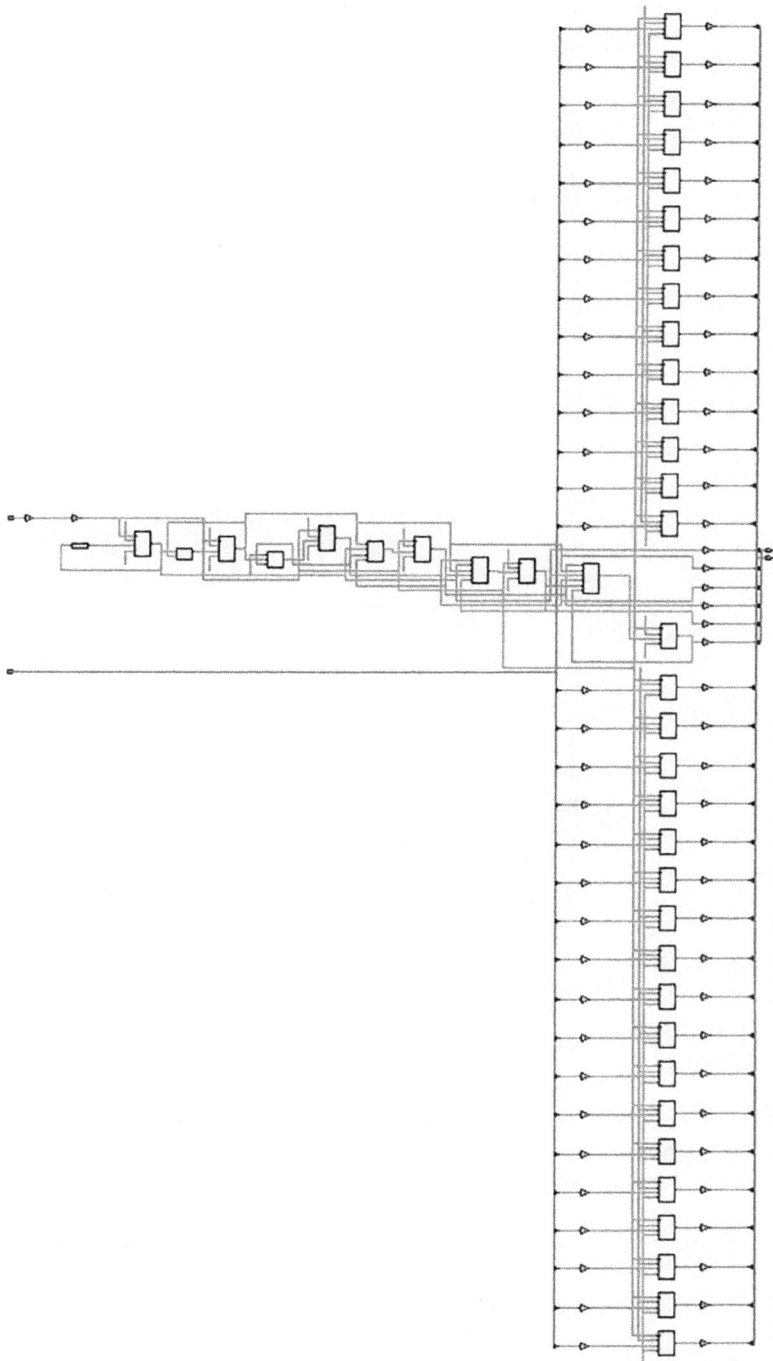

Figure 15.4 Technology schematic of packet counter.

Table 15.2 Thermal properties for the MOBILE DDR IO for Kintex-7

Frequency	JT (°C)	TM (°C)	ϑJA (°C/W)
100 MHz	25.4	74.6	1.9
300 MHz	25.4	74.6	1.9
500 MHz	25.5	74.5	1.9
1 GHz	25.6	74.4	1.9
5 GHz	26.7	73.3	1.9

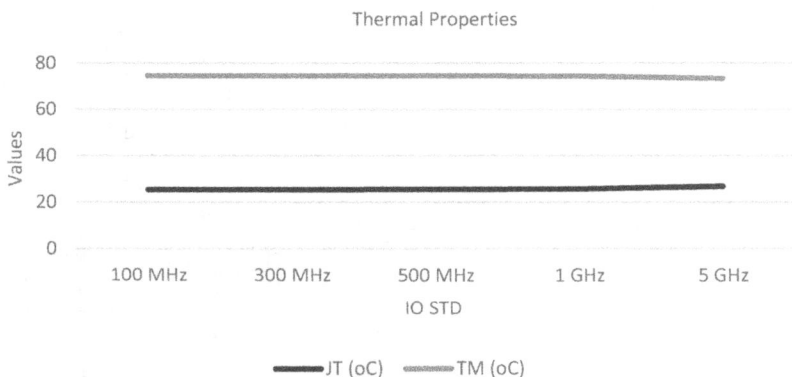

Figure 15.5 Thermal properties for the MOBILE DDR IO for Kintex-7.

The thermal properties for the MOBILE DDR IO for various frequencies are described in Table 15.2. The variation in thermal properties is described in Figure 15.5. From Table 15.2, it is observed that for the frequency of 100 and 300 MHz, the thermal properties remain the same. The change in JT and TM only observed for 500 MHz to 5 GHz frequency. ϑJA remains constant to 1.9 for all frequency values.

15.4 POWER ANALYSIS

In the current state of the globe, the power consumption of gadgets and information technology poses the greatest threat. The entire planet is experiencing an energy deficit. Thus, the entire ecosystem collaborates to produce a low-power solution for the electronics and information technology industries. Reduced power consumption prolongs the device's life. In this part, we calculate the power consumption of a packet counter implementation on a Kintex-7 device using the MOBILE DDR IO standard. The gadget's total power consumption is the sum of its dynamic and static power consumption [14–16]. This is expressed mathematically in Equation 15.1:

$$TP = SP + DP \qquad (15.1)$$

where

TP = Total Power
SP = Static Power
DP = Dynamic Power

The device SP is the total of the device's Clocks (clk), IO, Logic, and Signa (S/G) power, while the DP is the device's leakage power (LP).

15.4.1 Power analysis for 100 MHz frequency

When the frequency is tuned to 100 MHz value, the TPC of the device calculated is 0.206 W, which is the summation of DP (0.124 W; 60% of TPC) and SP (0.081 W; 40% of TPC). The TPC for 100 MHz is illustrated in Figure 15.6.

The DP is the total of IO, logic, and S/G power, which are, respectively, 0.124, <0.001, and <0.001. The SP is 0.081 W. The on-chips power for 100 MHz frequency is shown in Figure 15.7.

15.4.2 Power analysis for 300 MHz frequency

When the frequency is tuned to 300 MHz for matching the impedance of the device, the TP consumption of the device observed is 0.234 W. The device SP is 0.082 W which is 35% of TP, and DP is 0.153 W which is 65% of TP consumption. The power consumption of the packet counter for 300 MHz is illustrated in Figure 15.8. The graphical representation of TPC is described in Figure 15.9.

15.4.3 Power analysis for 500 MHz frequency

As the frequency of operation is increased to the value of 500 MHz, the TP of the device measured is 0.263 W, which is the cumulative sum of SP

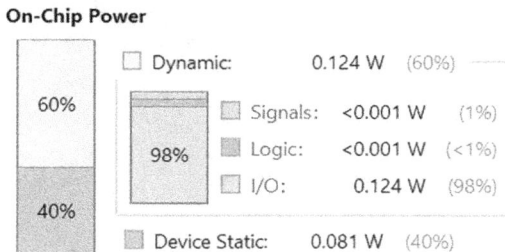

On-Chip Power

60%	Dynamic: 0.124 W (60%)
	Signals: <0.001 W (1%)
98%	Logic: <0.001 W (<1%)
	I/O: 0.124 W (98%)
40%	Device Static: 0.081 W (40%)

Figure 15.6 TPC for MOBILE DDR IO at 100 MHz.

Figure 15.7 Representation of on-chips power for 100 MHz.

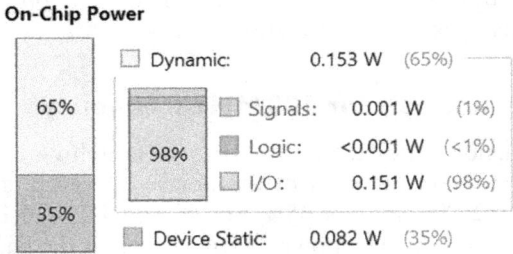

Figure 15.8 TPC for MOBILE DDR IO at 300 MHz.

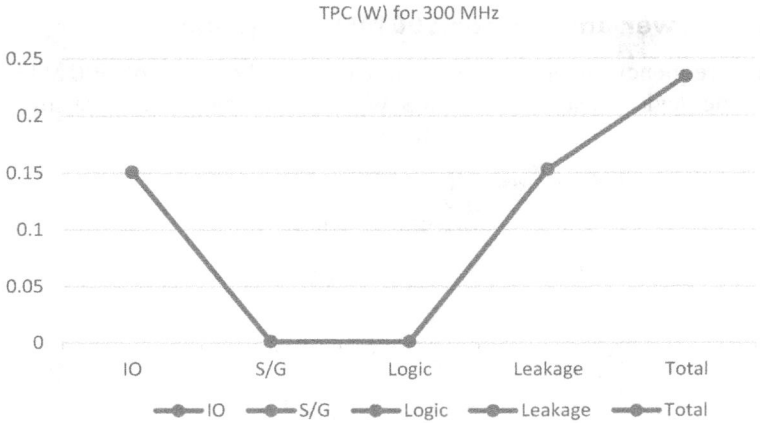

Figure 15.9 Representation of on-chips power for 300 MHz.

(0.082 W); 31% of the TP and DP (0.182 W); and 69% of the TP. The DP is the total of S/G, logic, and IO power, which are 0.002, <0.001, and 0.179 W, respectively. The representation of the total on-chips power is shown in Figure 15.10, and the graphical representation is described in Figure 15.11.

15.4.4 Power analysis for 1 GHz frequency

When the frequency is tuned to 1 GHz for the impedance matching, it is observed that the TPC is 0.335 W, which is the summation of DP (0.254 W) and SP is (0.082 W). The TP consumption for 1 GHz is illustrated in Figure 15.12.

We know that the DP is the summation of IO, S/G, and logic power. The IO, S/G, and logic power are 0.249, 0.004, and 0.001 W, respectively, for 1 GHz frequency. The graphical representation of on-chips power of the device is shown in Figure 15.13.

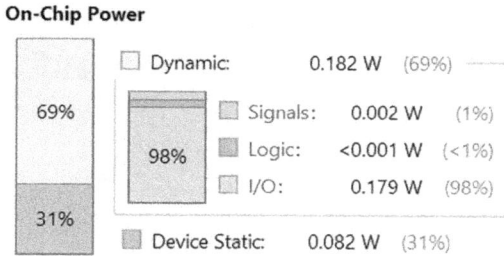

Figure 15.10 TPC of MOBILE DDR IO at 500 MHz frequency.

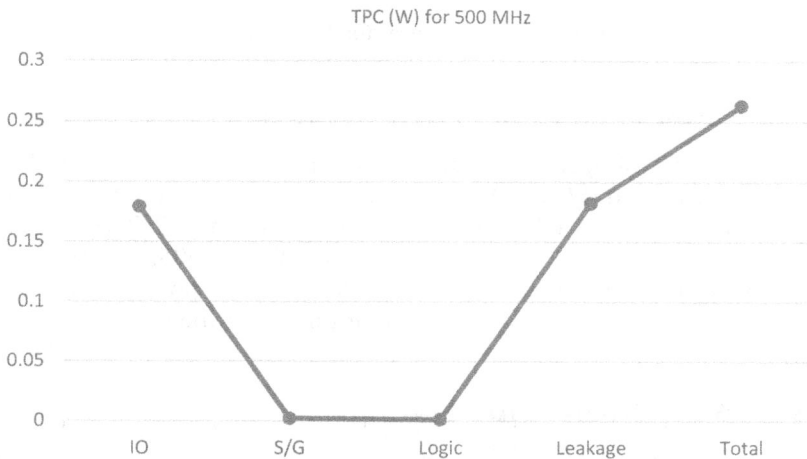

Figure 15.11 Representation of on-chips power for 500 MHz.

On-Chip Power

Dynamic:	0.254 W (76%)
Signals:	0.004 W (2%)
Logic:	0.001 W (1%)
I/O:	0.249 W (97%)
Device Static:	0.082 W (24%)

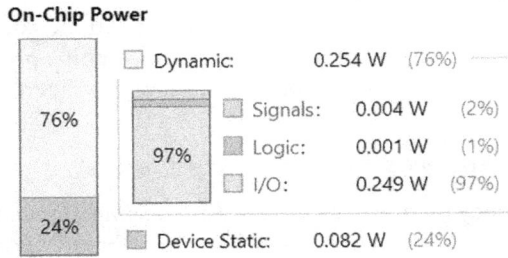

Figure 15.12 TP consumption for MOBILE DDR IO at 1 GHz frequency.

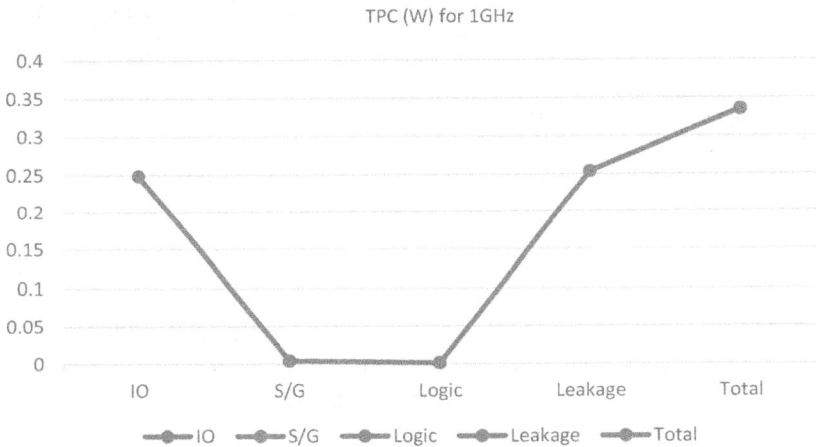

TPC (W) for 1GHz

Figure 15.13 Representation of on-chips power for 1 GHz.

15.4.5 Power analysis for 5 GHz frequency

As the frequency of operation is increased to the value of 5 GHz, the TP of the device measured is 0.913 W, which is the cumulative sum of SP (0.083 W); 9% of the TP and DP (0.830 W); and 91% of the TP. The DP is the total of S/G, logic, and IO power which are 0.021, 0.005, and 0.804 W, respectively. The representation of the total on-chips power is shown in Figure 15.14, and the graphical representation is described in Figure 15.15.

15.5 OBSERVATION AND ANALYSIS

As the frequency of operation for the device increases, the TPC for the MOBILE DDR IO increases, as shown in Section 15.4. The variation in TP is observed due to the variation in SP and DP of the FPGA device. This

On-Chip Power

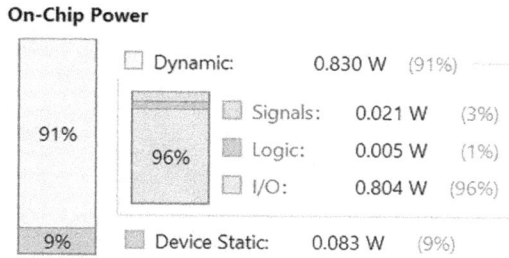

Figure 15.14 TP consumption for MOBILE DDR IO at 5 GHz frequency.

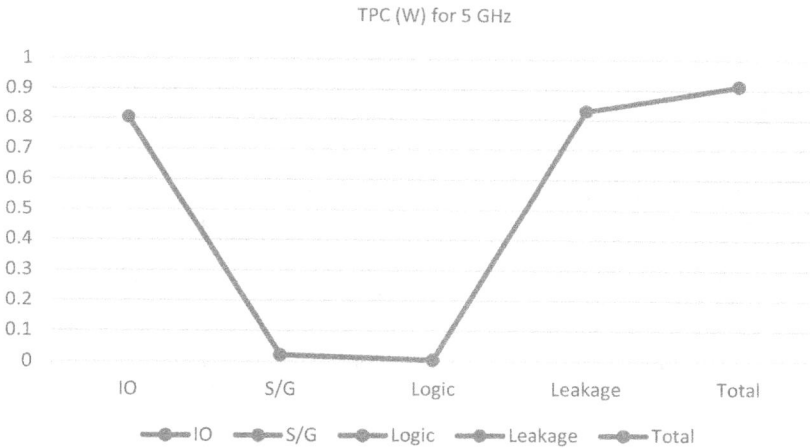

Figure 15.15 Representation of on-chips power for 5 GHz.

variation in SP and DP causes the TPC to increase. Figure 15.16 depicts the variation in TPC for various frequency values.

As the frequency of operation for MOBILE DDR IO increases, the TPC of the device also increases, as shown in Figure 15.16. The optimal MOBILE DDR IO frequency for this device is 100 MHz.

15.6 CONCLUSION

It has been observed that over the past few years, the world has been experiencing a severe energy and power shortage. The longevity of the earth's natural resources is uncertain. Eventually, it must vanish. In the context of such a massive problem, the concept of GC comes to mind. This chapter is an advancement in the GCC field. Using the Kintex-7 device, a power-efficient packet counter model is designed in this chapter. To optimize the power consumption of packet counter, the authors of this chapter have

TPC (W)

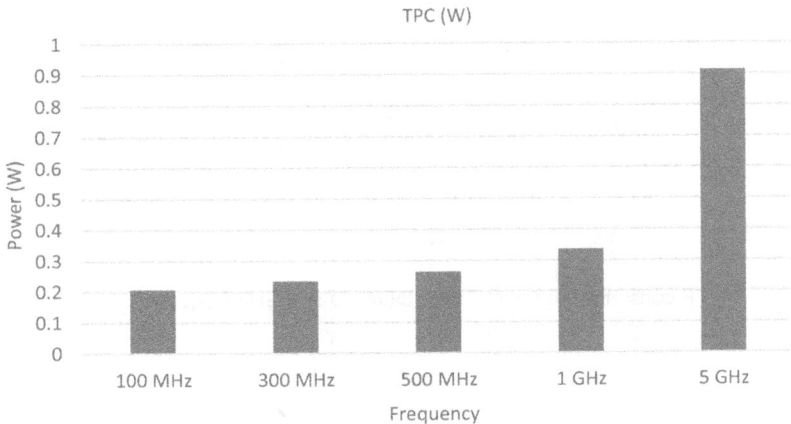

Figure 15.16 Variation in TPC for various frequency values.

employed the impedance matching technique, for which the MOBILE DDR IO standards have been used. In order to match the input and output impedance of the circuit, FPGA IO standards are utilized. This chapter examines the TPC of the device at various frequency values for the MOBILE DDR IO. The TPC of the device is observed to increase as the frequency of operation increases. The device provides the optimal amount of power at 100 MHz, while it consumes the maximum amount of power at 5 GHz.

In terms of its future applicability, the packet counter design can be implemented with other SoC-based FPGAs, and various power optimization techniques, such as capacitance scaling, voltage/current variation of the device, and clock gating, can be used. In addition, the FPGA design can be converted into an ASIC design for enhanced performance.

REFERENCES

1. Orlov, Sergei S., William Phillips, Eric Bjornson, Yuzuru Takashima, Padma Sundaram, Lambertus Hesselink, Robert Okas, Darren Kwan, and Raymond Snyder. "High-transfer-rate high-capacity holographic disk data-storage system." *Applied Optics* 43, no. 25 (2004): 4902–4914.
2. Rahmati, Ahmad, and Lin Zhong. "Context-for-wireless: Context-sensitive energy-efficient wireless data transfer." In *Proceedings of the 5th International Conference on Mobile Systems, Applications and Services*, San Juan Puerto, pp. 165–178. ACM Press, 2007.
3. Oruganti, S. K., S. H. Heo, H. Ma, and Franklin Bien. "Wireless energy transfer-based transceiver systems for power and/or high-data rate transmission through thick metal walls using sheet-like waveguides." *Electronics Letters* 50, no. 12 (2014): 886–888.

4. Jambunathan, K., E. Lai, M. A. Moss, and B. L. Button. "A review of heat transfer data for single circular jet impingement." *International Journal of Heat and Fluid Flow* 13, no. 2 (1992): 106–115.

5. Klein, Randall W., Michael A. Temple, and Michael J. Mendenhall. "Application of wavelet-based RF fingerprinting to enhance wireless network security." *Journal of Communications and Networks* 11, no. 6 (2009): 544–555.

6. Sinha, Preeti, V. K. Jha, Amit Kumar Rai, and Bharat Bhushan. "Security vulnerabilities, attacks and countermeasures in wireless sensor networks at various layers of OSI reference model: A survey." In *2017 International Conference on Signal Processing and Communication (ICSPC)*, Limassol, Cyprus, pp. 288–293. IEEE, 2017.

7. Banerjee, Suman, Bobby Bhattacharjee, and Christopher Kommareddy. "Scalable application layer multicast." In *Proceedings of the 2002 Conference on Applications, Technologies, Architectures, and Protocols for Computer Communications*, Pittsburg, PA, pp. 205–217. ACM Press, 2002.

8. Gupta, Isha, Swati Singh Garima, Harpreet Kaur, Deepshikha Bhatt, and Aamir Vohra. "28nm FPGA based power optimized UART design using HSTL I/O standards." *Indian Journal of Science and Technology* 8, no. 17 (2015): 1–6.

9. Kumar, Keshav, Amanpreet Kaur, S. N. Panda, and Bishwajeet Pandey. "Effect of different nano meter technology-based FPGA on energy efficient UART design." In *2018 8th International Conference on Communication Systems and Network Technologies (CSNT)*, India, pp. 1–4. IEEE, 2018.

10. Kumar, Keshav, Amanpreet Kaur, Bishwajeet Pandey, and S. N. Panda. "Low power UART design using different nanometer technology based FPGA." In *2018 8th International Conference on Communication Systems and Network Technologies (CSNT)*, India, pp. 1–3. IEEE, 2018.

11. Kumar, Keshav, Bishwajeet Pandey, Amit Kant Pandit, Y. A. Baker El-Ebiary, Salameh A. Mjlae, and Samer Bamansoor. "Design of low power transceiver on Spartan-3 and Spartan-6 FPGA." *International Journal of Innovative Technology and Exploring Engineering* 8, no. 12S2 (2019): 27–30.

12. Sandhu, Amanpreet, Vidhoytma Gandhi, Simranpreet Kaur, Surbhi Huria, Divjot Singh, and Wamika Goyal. "Thermally aware LVCMOS based low power universal asynchronous receiver transmitter design on FPGA." *Indian Journal of Science and Technology* 8, no. 20 (2015): 1–4.

13. Kumar, Abhishek, Bishwajeet Pandey, D. M. Akbar Hussain, Mohammad Atiqur Rahman, Vishal Jain, and Ayoub Bahanasse. "Low voltage complementary metal oxide semiconductor based energy efficient UART design on Spartan-6 FPGA." In *2019 11th International Conference on Computational Intelligence and Communication Networks (CICN)*, India, pp. 84–87. IEEE, 2019.

14. Pandey, Bishwajeet, and Ravikant Kumar. "Low voltage DCI based low power VLSI circuit implementation on FPGA." In *2013 IEEE Conference on Information & Communication Technologies*, India, pp. 128–131. IEEE, 2013.

15. Kaur, Harkinder, Harsh Sohal, and Jaiteg Singh. "Design and performance analysis of uart using Altera Quartus-II and Xilinx ISE 14.2." In *6th International Conference on Communication and Network Technologies*. India. 2016.

238 Green Communication with Field-programmable Gate

16. Kaur, Ravinder, Jagdish Kumar, Sumita Nagah, Bishwajeet Pandey, and Kavita Goswami. "IO Standard based low power memory design and implementation on FPGA." In *2015 2nd International Conference on Computing for Sustainable Global Development (INDIACom)*, pp. 1501–1505. IEEE, 2015.

Index

For Product Safety Concerns and Information please contact our EU
representative GPSR@taylorandfrancis.com
Taylor & Francis Verlag GmbH, Kaufingerstraße 24, 80331 München, Germany

www.ingramcontent.com/pod-product-compliance
Lightning Source LLC
Chambersburg PA
CBHW060357220326
41598CB00023B/2949